Chemistry as a Second Language: Chemical Education in a Globalized Society

ACS SYMPOSIUM SERIES **1049**

Chemistry as a Second Language: Chemical Education in a Globalized Society

Charity Flener Lovitt, Editor
Redmond, Washington

Paul Kelter, Editor
Northern Illinois University

Sponsored by the
ACS Division of Chemical Education

American Chemical Society, Washington, DC

Library of Congress Cataloging-in-Publication Data

Chemistry as a second language : chemical education in a globalized society /
Charity Flener Lovitt, editor, Paul Kelter, editor.
p. cm. -- (ACS symposium series ; 1049)
"Sponsored by the ACS Division of Chemical Education."
"This book evolved from an August 2009 symposium at the 238th annual meeting of the American Chemical Society in Washington, DC"--Pref.
Includes bibliographical references and index.
ISBN 978-0-8412-2590-9 (alk. paper)
1. Chemistry--Study and teaching--Congresses. 2. Education and globalization--Congresses. I. Lovitt, Charity Flener. II. Kelter, Paul B. III. American Chemical Society. Meeting (238th : 2009 : Washington, DC) IV. American Chemical Society. Divison of Chemical Education.
QD40.C453 2010
540.71--dc22
2010039488

The paper used in this publication meets the minimum requirements of American National Standard for Information Sciences—Permanence of Paper for Printed Library Materials, ANSI Z39.48n1984.

Distributed by Oxford University Press

Foreword

The ACS Symposium Series was first published in 1974 to provide a mechanism for publishing symposia quickly in book form. The purpose of the series is to publish timely, comprehensive books developed from the ACS sponsored symposia based on current scientific research. Occasionally, books are developed from symposia sponsored by other organizations when the topic is of keen interest to the chemistry audience.

Before agreeing to publish a book, the proposed table of contents is reviewed for appropriate and comprehensive coverage and for interest to the audience. Some papers may be excluded to better focus the book; others may be added to provide comprehensiveness. When appropriate, overview or introductory chapters are added. Drafts of chapters are peer-reviewed prior to final acceptance or rejection, and manuscripts are prepared in camera-ready format.

As a rule, only original research papers and original review papers are included in the volumes. Verbatim reproductions of previous published papers are not accepted.

ACS Books Department

Contents

Indexes

Empathy Is Global

In 1959, in my second year in graduate school in Harvard, I got it in my head to go to the Soviet Union on a one-year graduate student exchange. The scientific impetus came from some lectures Michael Kasha had given at Harvard, telling us of the important work of A.N. Terenin on the triplet state and A.S. Davydov on molecular excitons. But underneath I think I was still struggling with the question, did I really want to be a chemist. Halfway through a Ph.D. in chemistry!

I went to Moscow State University for a year in 1960-61. Everyone was against it. Harvard first - who goes abroad in the middle of a Ph.D? My mother too -- "They will draft you into the Soviet Army!" I was born in Ukraine, and we had come to America only ten years before. We (I was newly married; my wife Eva had just come from Sweden) went. And we experienced much of what the graduate students abroad describe so well in this book. It was not easy to live in Moscow – in midwinter the food stores had only cabbage, potatoes and onions. And tinned fish. But I have never regretted that year. On reflection, it provided a remarkable mix of cultural experience, personal growth at a critical age (I was 23), and lessons of empathy that have stood me well in my subsequent career as a researcher and teacher.

Let me explain the empathy. It has been my fortune to be put several times as a child or young person in the position of not knowing a language and a culture. I've mentioned two -- as an immigrant to the US from war-torn Europe, at age 11, the only English in my head from a year in school in Munich. And second, that stay in Moscow 11 years later. In each case I was an outsider –first a listener and a watcher, and then forced to act – to write that sixth grade paper on José de San Martin, to buy a bus ticket in Moscow. I think that experience, together with teaching introductory chemistry, helped me become a better theoretician. From standing outside, from being sensitive to the fact that I did not understand, I drew the conclusion that things were different for the people I was watching. Listening with empathy, thinking all the time about what is going on in the mind of the learner (or the reader of my paper), in time helped me shape effective explanations.

A kid (or an exchange scholar) in a different country thinks, tries to figure out how and why people do things. The smallest task -- that bus ticket -- is fraught with uncertainty as to process, a cultural setting, and language problems for the immigrant and visitor. Or for someone trying to locate distilled water in an African country.

What I learned from my year in Moscow was cultural empathy – that things can be done differently from the way I was used to, that one needed to understand the way common human physiology and root emotions were transformed by language, culture, and the political setting. That there was a reason, only seven years after Stalin's death, as to why we were never invited into a Russian home. I certainly became more sympathetic to the experience of an immigrant in the US (how quickly one forgets that one oneself was once an immigrant!). But also, as I taught introductory chemistry, I became more sensitive to the cultural difference that chemistry represents for a new learner, that learning chemistry is different from learning mathematics or biology.

Chemistry is a culture, and chemical thinking is a language. Cultural empathy is a thread that runs through the essays in this book. It brings together the experiences of graduate students, professors, and the NGO chemist active in a developing country. It's fascinating to read of the struggles to teach chemistry in South Africa, Afghanistan and the Iberoamerican world; it's not easy even back home in Italy. Or to read of being a graduate student in Germany or Slovakia, or starting pharmaceutical production in Cameroon. These accounts tell inspiring and amusing personal stories. They give very practical advice, and, at the same time, and much more broadly, testify to the desire and necessity of reaching across cultures, and trying to understand. With empathy.

Roald Hoffmann

Acknowledgments

The editors thank all of the authors for preparing interesting and enlightening chapters. Their hard work and experiences led to chapters that highlight the best aspects and greatest challenges of multicultural chemistry education. We also thank Taka Shimizu, Kwansima Quansah, Lucas Ducati, Erin C. Boone, Lina Chen, and Carlos Castro-Acuña for help in the translation and design of the cover art for this book, and Roald Hoffman for his encouragement and support via his preface. Additionally, we extend our sincere gratitude for the hard work and dedication of the editorial staff at ACS books, notably Tim Marney for his timely responses to our many questions, and Sherry Weisgarber for editorial work.

CFL sends thanks and love to AWL for encouragement during the editorial process. PBK sends love and thanks to BJK.

Always.

Chapter 1

Chemistry as a Second Language: The Effect of Globalization on Chemical Education

Charity Flener Lovitt*[,1] and Paul Kelter[2]

[1]University of North Texas, Department of Chemistry, 1508 W. Mulberry St., Denton, TX 76201
[2]Department of Teaching and Learning, Northern Illinois University, DeKalb, IL 60115
***chariteach@gmail.com**

Collaborations between scientists often transcend borders and cultural differences. The fundamental nature of science allows scientists to communicate using knowledge of their field but the institutions that support them are often hindered by financial and cultural barriers. As a result, science suffers. This book evolved from an August 2009 symposium at the 238th annual meeting of the American Chemical Society in Washington, DC. Its focus is on chemistry students and professors interested in developing a global approach to teaching chemistry, by participating in an international exchange program or incorporating culturally inclusive techniques into their classroom. The book has three broad themes; education research with a globalized perspective, experiences of teaching and learning in different countries, and organizations that support a global view of chemical education and chemistry. Here are the authors and an overview of their stories.

Chemical Education Research Perspectives

Liliana Mammino: Liliana Mammino is Professor of Chemistry at the University of Venda in a rural area of northeastern South Africa. Prof. Mammino is, in so many ways, a true Renaissance person of the rarest kind. Brought up in Italy, she spent many years in other countries, earning her Ph.D. in Russia at the University of Moscow and teaching in several countries in Africa before eventually beginning her position at Venda, where she's taught first-year and physical chemistry, and has done research in physical chemistry, for the past 13 years. Prof. Mammino's experiences have given her an unusually broad view of the world, and her fluency in four languages gives her a special - perhaps unique - insight into the impact of the mother tongue of students and their ability to learn chemistry. Her chapter, a true work of scholarship on several levels, explores the limitations to learning when chemistry instruction is given in a language other than the mother tongue. Citing scores of examples from her own experience, as well as a wide range of references, Prof. Mammino diagnoses the student misconceptions that occur under these circumstances. Examples of student statements include, "The *elements* that are listed above *are spontaneous*, based on the observations." "The *entropy* of the ice *is a perfect crystal*." "The above two equation / chemical reactions can be utilized in a galvanic cell, since *they* can undergo the redox reaction." "When T = 40 °C, the *temperature is noticeable*." She considers homophones ("same sound" English words that are interchanged, thus unintentionally changing the meaning), incorrect subject/verb coupling, omission of key words, difficulties in the use of prepositions, expressing comparisons, and others. Language-based reluctance to participate in discussions is but one outcome of many that result from students being taught in a second language. Prof. Mammino makes a convincing case for teaching in the mother tongue while students are in their formative years of learning science, and teaching also in other languages (notably English, the current international language of science communication) when the student is ready.

Liberato Cardellini: Professor Liberato Cardellini teaches and does research at Università Politecnica delle Marche in Ancona, Italy, looking out on the Adriatic Sea on the Eastern shore of the country. Some years ago, Prof. Cardellini began to reflect upon his teaching, deciding to become a better teacher by looking at the cognitive processes involved in teaching and learning. Engaging scholars worldwide, he learned about the nature of memory and its relationship to the traditional lecture, which he found to be a most unsatisfactory way of constructing knowledge. Rather, he found that the, "passive, non-thinking, information-receiving role" is unsuitable for learning. In this chapter, Prof. Cardellini considers the interaction between the inner mind and the outer way in which chemistry is, and can be, taught. He discusses the way experts think about a problem vs. the thinking of novices. He writes, "While the experts spend time in qualitative analysis of the problem, novices start with writing equations. Experts also tend to categorize the problem according to the laws of physics, while students categorize the problem according to some superficial entities and descriptions contained in the text of the problem. While the expert generates a physical representation of the problem, the novice often uses a process of direct syntactic translation." He then focuses on problem- solving in chemistry, writing, "it has been shown that the possession of chemical knowledge and the knowledge of strategies and skills are not sufficient to solve a problem if confidence arising from previous experiences of successful problem-solving is missing." He writes, "...the cognitive structures of good problem-solvers are more complex and contain more associations than those of poor problem-solvers. The strength of links among different concepts seems important in determining problem-solving behavior. It was also revealed that the deficiencies in the cognitive structures of poor problem-solvers appear predominantly for abstract concepts." He then goes on to describe best practices in problem-solving, including cooperative learning groups and the impact of teacher-based attributes. His chapter ends by considering the impact of these teaching methods on his students in Italy.

Students Who Studied Abroad

Markita Landry: Markita Landry is a well-traveled graduate student. As a child of a mixed cultural family, she understood the joys of international travel as she visited family in Bolivia at a very young age. "Cultural differences became apparent" for even a four-year old. She completed her undergraduate studies in chemistry and physics at the University of North Carolina at Chapel Hill where she learned all about basketball. She then enrolled as a graduate student at the University of Illinois at Urbana-Champaign, where she researches at the interface of chemistry, biology, and physics. While a graduate student at Illinois, Ms. Landry was selected as an East Asian Pacific Summer Institute Fellow to Japan and a US representative for the meeting of Nobel Laureates in Lindau, Germany. Her chapter discusses her experiences in the 10-week-long summer institute in Japan. The first third of her chapter is "A How-to Guide for the Aspiring Study Abroad Graduate Student," in which she lists the steps that a graduate student must take in order to prepare for, and successfully complete, a study abroad. She then details how international experiences lead to increased scientific productivity, where she discovered that although the science may be universal, "the *manner* in which these scientific questions are taught, learned, and researched varies greatly from laboratory to laboratory, and varies even more so from culture to culture." Her chapter closes with a discussion of the barriers that exist to cross-cultural exchange, in particular barriers "imposed on the scientific community by a country's economic or political standings (that) can greatly stymie scientific progress." Her experiences show that "multi-faceted problems require versatile solutions" and that international exchanges can be used to develop innovative research.

Charity Flener-Lovitt: Dr. Flener-Lovitt recently completed her PhD studies at the University of Illinois at Urbana-Champaign. Unlike the typical graduate student, localized into one field and one group, she spent her graduate career delocalized into chemical education, organometallic chemistry, and computational chemistry, which led to work in research groups in Illinois, Texas, Slovakia, and Germany. Dr. Flener-Lovitt first learned about chemistry abroad while spending the summer before graduate school working for a non-governmental organization (NGO) in Cambodia. The primary task of the NGO was teaching health and hygiene in rural Cambodian schools, but she jumped at the chance to use her chemistry background to test arsenic concentrations in local wells. She performed chemistry research in a primitive laboratory setting, where her labmates were chickens, dogs, and ants in a 5 foot tall ant hill. In graduate school, she earned the chance to research in Central Europe as a Central European Summer Research Institute Fellow. After spending one summer in Europe, she applied for and received a US State Department Fulbright Fellowship to Germany. In these settings, she learned that chemistry is a universal language. Her chapter details myths that prevent graduate students from traveling abroad and details the application process for short-term and long-term study abroad fellowships. She then discusses the impact of the study abroad on her graduate career on her professional and personal life. Her chapter ends with a list of tips for graduate students that may decide to apply for study abroad fellowships.

LeighAnn Jordan: LeighAnn Jordan is currently a graduate student at Michigan State University. While she was undergraduate at Westminster College in Pennsylvania, she participated in a summer study abroad experience in Germany. Her experience there "opened [her] eyes to the international community" and helped her discover that "research is truly an international effort, and should not be separated by language barriers and/or country borders." Her experience abroad helped confirm her choice to study biological chemistry with a basis in medicine, specifically so she could collaborate with professors in departments outside of chemistry and outside of the US. In addition to affirming her choice of career, her chapter discusses how her experience abroad led her to seek graduate schools away from home. Her chapter provides tips essential to undergraduates who may consider participating in a summer research experience outside of the United States.

Teaching in Diverse Cultures

COL (Ret.). Patricia Dooley, PhD: Patricia Dooley is on the faculty of Bard College at Simon's Rock in Massachusetts. This is, however, a second career for her. In her first career, she was a long-time member of the U.S. Army, rising to the rank of Colonel before retiring in 2008. While in the Army, COL (Ret.) Dooley served successfully in Asia, Europe and the United States. Her last overseas trip was to Afghanistan, where she served as a mentor and advisor to the National Military Academy of Afghanistan (NMAA) in the capital city of Kabul. Her experience helping to rebuild the University of Kabul's chemistry program is the focus of her chapter. In her abstract, she describes the country as, "...reviving itself after 27 years of occupation, civil war, and governance by the Taliban, and still combating an insurgency..." Her stories of making something out of nearly nothing attest to the struggles to build an intellectual life. In her text, she describes the conditions there, "While not in written or spoken language, there is universality in a flooded chemistry laboratory floor—especially in a building with no running water, lights, or electricity... Kabul University had no chemicals to spare. Their laboratories had been plundered of everything: windowpanes, light fixtures, shelving, drawers, plumbing, electrical outlets. Seeing the great losses this institution had endured made the conditions at NMAA [National Military Academy of Afganistan] look luxurious in comparison." Hers is a narrative of people working together across the barriers of geography, language, politics, and social customs to create the conditions for the people of Afghanistan to learn the international language - of chemistry.

Profs. M. Carlos-Acuña and Paul Kelter: These two long-time friends and colleagues have worked together, most often at a distance of 2700 km, in the service of chemistry education for nearly 20 years. They met at an international conference, and saw so much common ground that it showed how similar seemingly distant cultures can be. About six years ago, Castro-Acuña and Kelter decided to start an organization of college and university teachers dedicated to supporting those who sought to improve the teaching of first-year chemistry worldwide. Since then, the International Center for First-Year Undergraduate Chemistry Education (ICUC) has been the vehicle for a vibrant level of collaboration among hundreds of first-year chemistry teachers throughout the world. The ICUC (pronounced "E-Cook" in Latin America), has run conferences, led symposia, fomented research, development and friendships via sabbaticals, and encouraged publications. In the chapter, several Iberoamerican ICUC members discuss the impact of the organization. Founding Board member José Miguel Abraham, a professor at the Universidad de San Luis in, Argentina, notes, "I have always taught that chemistry can contribute to the preservation and/or recovery of the environment in its natural, social and human aspects...To be involved with the ICUC since its beginnings is something that has given me a lot of satisfaction and the opportunity to increase my knowledge and to share my ideas with teachers from all around the world." Amalia Torrealba, from the Universidad Central de Venezuela in Caracas. Dr. Torrealba notes, "The fortress of this association is the integration of a considerable number of teachers from several countries, which allow us to examine educational problems from different perspectives, and to generate ways to solve them." The ICUC has also led many teachers to look in a deeper way than ever in their careers at their teaching philosophy and practice. This chapter describes the growth of the organization and the challenges (most notably financial) to its continued vitality.

Implications of the Globalization of Science

Dr. Rolande Hodel: Rolande Hodel is the founder of AIDSfreeAFRICA, a non-profit organization that seeks to establish sustainable pharmaceutical drug production in Sub-Saharan Africa. She regularly travels between her home on the east coast of the US and Cameroon. Her chapter provides an interesting perspective on teaching science in developing countries. She details the need for training chemists in developing countries so life-saving drugs can be developed in-country. Globalization of science has lowered the barrier to drug production, but the lack of training and resources prevent medical start-ups from succeeding in developing countries. Her chapter provides compelling reasons for US scientists to become more active in training and learning more about science in developing countries.

Organizations That Fund Study and Teaching Abroad

* *Humboldt Foundation Fellowships (Germany)*
  http://www.humboldt-foundation.de/web/start.html
* *The German Academic Exchange Service (DAAD)*
  http://www.daad.org/
* *Institute of International Education*
  www.iie.org
* *Whitaker International Scholars Program*
  http://www.whitakerawards.org/
* *Kaust Discovery Scholarships (Saudi Arabia)*
  http://www.kaust.edu.sa/
* *Fulbright Fellowships*
  http://www.iie.org/fulbright/ (Students)
  http://www.cies.org/ (Professors and Professionals)
* *US National Science Foundation - Developing Global Scientists and Engineers*
  http://www.nsf.gov/funding/
  pgm_summ.jsp?pims_id=12831&org=OISE&from=home

We would like to thank the Division of Chemical Education of the American Chemical Society for the opportunity to organize the symposium and ACS books for the invitation to prepare this book. We are grateful for the contributions for our distinguished authors and for their insight into science education beyond US borders.

Chapter 2

The Mother Tongue as a Fundamental Key to the Mastering of Chemistry Language

Liliana Mammino*

Department of Chemistry, University of Venda, P/bag X5050, Thohoyandou 0950, South Africa
***E-mail: sasdestria@yahoo.com**

Language mastering plays fundamental roles in the development of scientific thought and, consequently, in learners' acquisition of scientific knowledge. The language mastering level and the internalization depth of word-concept and expression-concept correspondences within the mother tongue are unparalleled by any other language that a person may use, making the mother tongue the optimal vehicle for students' familiarization with the concepts and methods of science and for the development of skills essential to such familiarization, like visual literacy, logical abilities and abstract thinking. *This chapter provides extensive documentation on the impacts of using a language different from the mother tongue to approach chemistry, through the analysis of the difficulties encountered by tertiary level chemistry students in second-language disadvantaged contexts.* The results stress the importance of utilizing the mother tongue to approach chemistry, at least until the student acquires sufficient familiarity with the chemistry discourse and simultaneously develops adequate mastering of relevant skills. This acquisition will constitute a solid foundation enabling clear identification of the chemistry discourse when utilizing other languages, so that the use of other languages can expand the range of communication possibilities and add the benefits ensuing from the different reflection-perspectives inherent in using different expression tools.

Introduction: The Questions, the Context and the Approaches

Language and Science; Language and Chemistry

Language is the fundamental instrument for the development and communication of thoughts. As such, it has an essential role in the development of science and, simultaneously, the needs arising from new developments in science imply new developments in language: "When an area of scientific thought is new, the interpretative role of language is central. New ways of *seeing* what is going on are closely connected with new ways of talking about it" (*1*). To acquaint students with this fundamental aspect of the nature of science, to foster creative scientific thinking, science teachers should thus simultaneously be language teachers (*1*), acquainting students with a given science as a discourse through integrated acquisition of new terms and new thoughts, and of the ability to express the new thoughts.

The *language of science* is not just terminology. Communication or thought generation require much more than technical terms (the names of the entities and/or phenomena that are the objects of a given scientific discourse), because they depend on the words linking the technical terms to build a meaning, and these are the common words (verbs, adjectives, prepositions, logical connectives, etc.) pertaining to the language utilized (*2*, *3*): technical terms are thus *immersed* in a *sea* of common words that constitute the backbone of the communication. Knowing the meaning and roles of these common words, being able to understand what they communicate (on reading or listening) and to use them so as to communicate a wanted meaning (on speaking or writing) become essential instruments to ensure correctness and clarity in any form of communication (*2*, *3*). In particular, rigorous wording usage by the teacher enhances the quality of explanations and helps prevent confusion and misconceptions (*4–8*). The inevitable general inference is that language mastering is the key to science learning as well as to creativity in the sciences. This, together with the acknowledgment of the paramount internalization depth of word-concept and expression-concept correspondences within the mother tongue, points to the essential role of the mother tongue as the natural ground to develop language mastering up to the highest sophistication levels and, in particular, up to the levels that are needed for science communication and science learning. The essential role of approaching science through the mother tongue, to learn to recognize science as a language – a recognition that, once ignited, can naturally extend beyond the mother tongue to any other language that the student may use – constitutes the major focus of attention in this chapter.

Chemistry and chemistry education are particularly apt to highlight language-related aspects in science teaching/learning, for the same reasons for which chemistry can be viewed as an ideal area for *language-of-science* education (*9*). The simultaneous extensive presence of descriptions through language and through mathematics, with two intertwining description levels (macroscopic and microscopic), the use of a symbols system that is probably the most extensive and articulated in the sciences, and the continuous interplay between observation and interpretation, demand substantial language-mastering sophistication-level to be applied to a range of investigation domains whose diversity is probably unmatched by other sciences. The known difficulties students encounter at approaching chemistry can often be traced to difficulties at understanding texts (*10*) – with reference to both written and oral communication – and this, in turn, often depends on inadequacies in the language-mastering sophistication-level (*11*). Interventions aimed at enhancing language-mastering can thus be viewed as perspective relevant components of chemistry education.

Background Information on the Context and the Analysis Approaches

The reflections presented in this chapter are based on over 20 years experience with teaching general chemistry and physical chemistry courses in Southern Africa – the last 12 years at the University of Venda (UNIVEN) in South Africa, where the teaching also included the process technology course (an introduction to chemical engineering, largely based on physical chemistry). UNIVEN is a particularly disadvantaged institution, combining the *historical disadvantage* of being a *Historically Black University* (HBU – a university that was "for blacks only" during the apartheid period, which, according to the political criteria of those times, implied both poor resources and poor educational approaches, aimed at ensuring that black students would not have a chance to excel) and the socio-economic disadvantages common to poor rural contexts (*12*). In particular, the country-wide general scarcity of qualified secondary school science teachers affects rural areas more extensively, resulting in the serious underpreparedness of most students entering UNIVEN.

Long-dated personal interest in the *language of science* and in the ensuing pedagogical implications (*2*, *13*) provided a background attitude of specific attention to language aspects. On the other hand, having grown and been educated in a mother-tongue-instructional context (Italy), it took me some time to arrive at the realization that not having approached science through the mother tongue was the major single cause of the difficulties experienced by students to understanding, learning and expressing chemistry. The absence of mother-tongue-based instruction is actually recognized as a major obstacle to development in Sub-Saharan Africa because of its heavy impact on the acquisition of knowledge and expertise (*14–17*) – an impact that becomes particularly serious for the science disciplines (*18*, *19*). The main question for me as a chemistry lecturer became the identification of the details of how this factor affects students' approach to chemistry and the extent (or even the very possibility) of their understanding it.

An analysis of students' answers from the combined (and, as much as possible, integrated) points of view of language aspects and chemistry aspects was selected as the principal investigation tool. The analysis (still in progress) has brought extensive information: it has enabled the identification of categories of errors that would never (or very rarely) occur within the mother tongue (*20–24*) and the identification of limitations ensuing from poor language-mastering, from the difficulties with logical frameworks to the near annihilation of the benefits that could potentially derive from the use of alternative communication tools like visualization, because language-mastering inadequacies prevented the development of the ability to use them. Although my mother-tongue mental reference is not English, general evidence from language usages – like the fact that people do not confuse words differing simply by tones or accents, as long as they pertain to the mother tongue – are considered sufficient justification for the assumption that confusions of the types highlighted in (*21–24*) would not (or very rarely) occur within the mother tongue. As for the features concerning logical relationships, logical frameworks and other aspects of the science discourse, they transcend individual languages and, therefore, the analysis-references are practically language-independent and the analysis itself highlights the extent to which students are able to recognize and/or express them. Classroom interactions have played crucial complementary roles as sources of diagnostic information: besides providing verifications of the hypotheses based on the analysis of students' writings, they have had primary functions in highlighting the heavy impacts of language-related difficulties on the development of other abilities that are fundamental for understanding and learning chemistry, first of all visual literacy and the ability to identify and follow logical frameworks.

The diagnostic part from the observations up to the 2008 academic year (*25*) included is documented in previous works (*12*, *21–24*). The reflections stemming from the observations, analyses and diagnoses go beyond the documentary evidence, as they tackle the question of how language determines a young person's approach to science and how, in turn, this relates to the acknowledged universal (or cross-languages) characters of the *language of science* (*3*). The conclusion is that the mother tongue is essential to the first familiarization with the nature of science and the stimulation of scientific curiosity, and to the acquisition of essential abilities, from visual literacy to logical thinking and abstract thinking. Once these foundations are acquired (in the pristine meaning of the term, i.e., the student *comes to own* them), then the universal characters of the *language of science*, or of science as a language, become accessible, and reading (or listening to) science communication through other languages may stimulate new reflections and new insights, enhancing understanding instead of being an obstacle to it.

Significance of the Information from Disadvantaged Contexts

Since the reflections presented here are mostly based on observations in disadvantaged contexts, it appears natural to ask whether the significance of the information from a disadvantaged context can extend beyond that specific context, or beyond the ensemble of similarly disadvantaged contexts.

The information from a disadvantaged context depicts generally poorer or even extreme scenarios (depending on the extent and impact of disadvantage-generating factors). By doing so, it offers evidence of the ultimate consequences:

- of features that might be present in other contexts to a lesser extent (so that their impact is not as evident), or
- of tendencies whose potential impacts might be currently overlooked, or
- of options or policies that might be taken into consideration for possible adoption without sufficient realization of their disadvantage-generating potentialities.

For the issue of the language of instruction for science subjects, Sub-Saharan Africa, in particular, offers univocal evidence of the importance of mother tongue instruction, by showing the impacts of not adopting it. This documentary evidence is relevant for other contexts as, in a number of contexts, the role of English as the current *lingua franca* for scientific communication is prompting the question of whether it might be better to teach science in English (instead of the mother tongue) from the beginning of a pupil's formal education in science. The evidence from Sub-Saharan Africa (that can be rightly considered as massive experimental evidence) gives a univocal answer: the mother tongue has fundamental roles that cannot be replaced. This also implies that the familiarization with the *lingua franca* – whose importance is undeniable – needs to be pursued through options other than the replacement of the mother tongue, because what would be diminished or lost on such replacement is the very contact with the nature and methods of science. In this way, the inferences from the analysis of the situation in disadvantaged contexts offer precious indications for other contexts.

Chemistry Students' Language Related Difficulties in Second-Language-Instruction Underprivileged Contexts

Diagnoses from the Analysis of Students' Written Works

Documenting Language-Related Difficulties from Chemistry Students' Works

Language-related difficulties are observed for all the aspects concerning language and its usage to express chemistry, from the selection of individual words to the coupling or association of relevant pairs or triplets of words (subject-verb, subject-verb-object, adjective-noun, etc.), the use of prepositions, the building of individual clauses and the use of logical connectives in the construction of complex sentences. The consideration of concrete examples is the only way to convey a tangible perception of the nature of the difficulties. The examples included in this chapter are all from students' written works (reports, tests, etc.) in the 2009 academic year (*25*), to respond to the goal of maximum updated-ness (as they are the most recent available). Ample selections of examples pertaining to previous academic years can be found in the works analyzing individual language-related issues (*22–24*) or their impacts within specific courses (*26*); their comparison with those included in the current work highlights the continuity and deepening of the language-related difficulties experienced by chemistry students.

The examples are selected considering errors for which the language component is clearly dominant. In many cases, it is difficult to untangle the language-related component and the chemistry/concept-related component in incorrect statements, although-language related difficulties constitute either significant components or the major cause of many incorrect statements, because of their impacts on conceptual understanding, on the acquisition of skills and on the development of attitudes. On the other hand, there are errors that can be ascribed to language-related causes with a good degree of probability, and these are the ones that are selected as documenting examples. For instance, no example of the very frequent confusion between *heat* and *temperature* is included in the selection, because this confusion is known to be mostly conceptual and spread through many contexts, although it is clear that, at UNIVEN and other second-language disadvantaged contexts, it also has significant language-related roots. Similarly, recurrent errors like those concerning the identification of the dependent and independent variables (in relationships, diagrams, experiments or discussions) are not included because the conceptual component is presumably dominant (although quite often the student's language mastering level is not adequate to enable him/her to really understand the difference between the two variables categories, i.e., to reach the needed conceptual understanding that is pre-requisite to correct expression).

The examples are reported as they were originally worded by the student, without correcting other errors that might be present, because this provides a more complete picture of the overall situation. The errors reported cannot be ascribed to typing mistakes because all students' works are handwritten. The examples are numbered progressively to facilitate references throughout the chapter and, when the error in question refers to one or two individual words, these words are written in italics for more immediate identification.

In considering the examples, it is necessary to constantly keep in mind that passive memorization is the students' generalized answer to language-related difficulties (*21*, *26*). Not being able to attain satisfactory understanding of the meaning of the sentences and descriptions that they read, students simply memorize (or try to memorize) them. The correctness of the regurgitation of passively memorized material depends solely on the memory abilities of the students, as the language-mastering degree of most students is not adequate to enable them to identify the meaning actually conveyed by the sentences that they have written, on proofreading, which prevents them from identifying errors.

Confusion Related to the Sound-Concept Correspondence

A type of confusion that would never occur in the mother tongue (*27*) concerns the sound-concept correspondence (*22*). Homophone (or nearly homophone) English words are often interchanged in students' writings. The interchange frequency has been increasing in recent years and extending to pairs of words whose homophone-ness is more distant than for the cases observed up to 4–5 years ago (and that would, e.g., not be perceived as homophone pairs by European non-native English speakers). In a number of cases, it is reasonable to assume that the student knows the meaning that he/she is trying to express and the language error can be viewed as not affecting the chemistry. It is the case of interchanges between *presence* and *presents, packed* and *parked, begging* and *beginning, species* and *spices, ones* and *once, course* and *cause, deep* and *dip, sea* and *scale, string* and *stir, to* and *two, convention* and *conversion, objection* and *objective, exist* and *exceed* (1–13) and many other pairs in which one of the terms is not normally utilized in the descriptions to which the given statement/sentence pertains:

1. The *presents* of intermolecular interaction affects the behavior of the real gas.
2. In a solid, molecules are *parked* together.
3. At the *begging* of the reaction the graph is constant.
4. When the *spices* in an anode is oxidized it releases electrons.
5. The atom are the *once* which deposited on the copper rod.
6. The energy is of *cause* quantized since there are boundary conditions.
7. *Deep* the copper rod in a solution containing copper sulphate.
8. Water boils at 100 C at *scale* level.
9. We use a *string* rod to mix the solution.
10. Electrolytic cell: *to* metal rod dipped in a solution.
11. The *convention* of solid to liquid is possible only if there is a heat supply.
12. The *objection* of this experiment is ...
13. We know that in this experiment the temperature does not *exist* 100 °C.

The teacher's good will can stretch to accept that a student writing *inaccuracy* actually meant *accuracy:*

14. The *inaccuracy* of measurements is affected by an impurity present.

In other cases, both interchanged terms belong to the description of the given issue and, therefore, it becomes more difficult to reach a reasonably founded idea of what the student might want to express, or how clear are his or her views. For instance, the terms *transform* and *transfer* are both relevant for the description of redox reactions (15; the issue is discussed in detail in (*22*)); the temperature increase on supplying heat to a system depends both on the *composition* of the system and on the *conditions* under which the heat is transferred (16); *orbit* and *orbital* pertain to different models of the atom (17):

15. Before the molecules of the reactants are *transferred* into the molecules of the products, the reactants must achieve the minimum energy.
16. The magnitude of the temperature increase depends on the *composition* under which heat is transferred.
17. In Bohr's model, the electron moves in *orbitals* around the nucleus.

Similarly, both *constant* and *coexistence* pertain to the description of the isotherms of a gas: there are parts of the curves corresponding to *constant* pressure and to the *coexistence* of the liquid and gas phases. The confusion between the two terms and the overall wording of statement 18 highlight a total failure to understand the concepts; the way in which *coexistence* is used suggests that the meaning of this terms remains unknown:

18. The vertical line correspond to the *coexistence* of the volume. The volume remains the same or is *constant* when the pressure is decreasing.

Even these examples, concerning a type of confusion that could be viewed as the most elementary language-related one, highlight a general problem of evaluating students' works: language-related difficulties become a *confounding variable* (*28*) complicating the assessment and often decreasing its reliability. It is difficult, often impossible, to untangle the language component and the chemistry component in a mistake, or to ascribe them relative weights, and, therefore, it becomes difficult to assess students' acquisition of chemical knowledge. Too often, the student is unable to show what he or she knows because of being unable to express it, and the accuracy and fairness of the assessment are compromised.

Doubts about the assessment accuracy are further enhanced by some experiments (*29*) conducted in another context (the National University of Lesotho): students who had written incorrect or meaningless statements were asked to explain their views on the given issues (the theme/s of the questions) through their mother tongue to somebody who could then translate their answer into English (*30*); in several cases, the translated answer corresponded to reasonable chemistry; further discussions highlighted the details of the language difficulties that had led to absurd or meaningless answers (often related to grammar and sentence-construction, but also to the selection of individual words, or to how to combine them to express the desired meaning).

Confusion Concerning the Association of Words with Specific Roles

Errors in the association/coupling of words with specific roles are very common, and often suggest incorrect chemistry. Incorrect coupling may concern the subject and the verb, like the use of *consist* (19, 20) or *occur* (21–24) in relation to the selected subjects, the identification of what *is noticeable* in the isotherms of a gas approaching its critical temperature (25, 26) or the identification of what can *melt*.

19. The *equations* for real gases *consist* of the parameters which depend on the chemical nature of gas.
20. The *reduced mass technique consists* of a particle of reduced mass μ and which moves around a stationary nucleus of infinite mass.
21. *Electrolytic cell* uses electric current to *occur*.
22. Cathode is a place where the *species* that is reduced can *occur*.
23. Objective of the experiment: to activate *non-spontaneous reagents to occur* by electric current.
24. *Reduction potential* is the potential that *occurs* at the reduction half-reaction.
25. At T = 40 °C, the *trend is noticeable*.
26. When T = 40 °C, the *temperature is noticeable*.
27. At T = 50 °C, *liquid melts* completely.
28. The *temperature* of the ice *was melting* after 420 s.

It may be interesting to analyze the mistakes in the previous statements more in detail. The verb *consist* introduces a complete description of what the subject is made of; thus, it cannot be used if the list is not complete (19, where the correct verb is *contain*) or if what follows is not a description of what the subject is made of (20, where correct forms could be *the reduced mass technique utilizes a model in which a particle..., in the reduced mass technique, we consider a particle....* or other forms consistent with the meaning that the technique replaces the actual system with the model described by the rest of the sentence and not implying that the technique itself is made of particles). The verb *occur* can be associated with events (processes, phenomena), not with objects, whether the objects pertain to the physical world, like the *electrolytic cell* (21), the *chemical species* (22) and the *reagents* (23), or to our models, like the *reduction potential* (24). Correct statements remaining as close as possible to the students' statements could be the following: *An electrolytic cell uses electric current to force a non-spontaneous redox reaction to occur* (21); *The cathode is the electrode where a species is reduced* (22); *Objective of the experiment: to force a non-spontaneous redox reaction to occur by using electric current* (23); *Reduction potential is the electrode potential associated with the reduction half-reaction* (24). It can also be noted that the error in case 23 is likely due to confusion between subjects (*reagents* in place of *reaction*) that the student might perceive as homophone. In a typical diagram of the isotherms of a gas (25, 26) what is *noticeable* for isotherms approaching the critical temperature is the appearance of *deviations from ideal behavior*; the term *trend* (25) is too generic to convey information on what happens for isotherms approaching the critical temperature, and *temperature* (26) is the variable characterizing each isotherm, so, it cannot become *noticeable* only for some of them (the term *noticeable* is present in the study material and it surfaces frequently in students' statements, but passive memorization does not help understand its meaning or role in the sentence). The subject of *melt* should be a *solid*, not a *liquid* (27) or *temperature* (28).

Oversimplifications and the Omission of Key Words

In a number of cases, what emerges as incorrect identification of the subject-verb pair results from oversimplification of a sentence structure, with omission of key words; e.g., writing *lower standard potential* in place of *the species with lower standard potential* (29), or *galvanic cell* in place of *the e.m.f. of a galvanic cell* (30), or *first order reaction* in place of *the half-life of a first-order reaction* (31):

29. *Lower standard potential is oxidized*, not higher standard potential.
30. *Galvanic cell depends* on temperature.
31. *First order reaction* does not depend on the initial concentration.

The omission of words that are essential for the meaning of the sentence is a very frequent phenomenon, and mostly depends on difficulties in formulating sentences with complex wording, even in one-clause sentences like the previous ones, often combined with the limitations associated with passive memorization (as memorization difficulties increase with wording complexity). The tendency toward wording oversimplification, combined with difficulties at relating subject and verb, quite often results in the use of un-defined subjects – pronouns whose reference is not identifiable from the context (32–34) or whose strictly grammatical reference-identification would yield a meaningless statement (35, suggesting that a reaction is a straight line, or 36, suggesting that redox reactions undergo redox reactions, or 37, suggesting reference to the probability of finding the wavefunction, instead of the probability of finding the particle):

32. This is because *it* involves transfer of electrons.
33. Cathode ==> the electrode in which *it* is reduced is called cathode.
34. The concentration are different so *it* move from low concentration to high concentration.
35. This is first order reaction. Because *it* is a straight line.
36. The above two equation / chemical reactions can be utilized in a galvanic cell, since *they* can undergo the redox reaction.
37. Wavefunction must be single-valued so that the probability of finding *it* is unambiguous; it must be a single value so that *it* exists.

That these errors are largely language-related is confirmed by attempts to prompt students to analyze the sentences, showing difficulties in identifying the literal meaning of the sentence: the same language-related difficulties that make it arduous for students to understand a written text (in books or other study material), make it arduous for them to proofread their own writings and perceive/recognize the meaning conveyed.

Difficulties in Expressing Qualities and Attributes

The most straightforward expression of qualitative aspects is through the «subject + verb *to be* + an adjective or a noun» combination. The ways in which students select and combine these words often shows difficulties at relating a certain quality to a certain entity, as illustrated by examples 38–43. Cases 38 and 39 ascribe the *spontaneous* qualification to subjects to which it cannot be applied: it cannot be applied to an object (like an element), nor to the *experiment* concept, because it cannot be spontaneous by its very methodological nature, since we are the ones who plan and organize an experiment. Case 39 likely pertains to the omission-of-key-terms tendency, as the correct subject would be *the reaction occurring in this experiment.* Case 40 refers to a widely spread difficulty concerning the understanding of chemical equilibrium, for which students tend to consider that all the concentrations should be equal at equilibrium (a misconception diagnosed in many contexts, not only underprivileged); but the literal meaning of the sentence would refer to the chemical nature of the species; random questions showed that students were not aware of its literal meaning: some of them meant *concentrations*, others were not sure about what should *be the same.* Case 41 ascribes to the reaction the quality (being negative) that pertains to its ΔG. Case 42 identifies a physical quantity with an object – an error that is partially related to poor familiarity with the foundations of the scientific method, resulting (among other things) in poor familiarity with the nature of different entities, like physical quantities, objects, processes, etc. and their roles in our descriptions; but language-related factors are dominant, as random questions showed that those students did not actually believe that entropy is a crystal, but they did not have a perception of the literal meaning of their statement. Similarly, statement 43 is unlikely in the mother tongue, because the deep internalization of the concept-term relationship (the association of mental images to words) in the mother tongue would prevent somebody from considering *ice* as a *gas*.

38. The *elements* that are listed above *are spontaneous*, based on the observations.
39. This *experiment* is *spontaneous*.
40. At equilibrium all *the species must be the same*.
41. The spontaneous *reaction* is always *negative*.
42. The *entropy* of the ice *is a perfect crystal*.
43. I will assume that *ice is ideal gas*.

Difficulties in the Use of Prepositions

Prepositions are the simplest logical connectives, with essential meaning-building roles in individual clauses. Within the mother tongue, their meaning is so deeply internalized that errors in their use are rare. However, such errors are very common in a disadvantaged second-language context, resulting in a serious loss of the meaning of sentences and, frequently, in difficulties in the further treatment of the concepts involved. Among the illustrative examples reported, sentences 44–47 pertain to descriptions, and the replacement of the expected preposition by an incorrect one (*of* in place of *on* in 44, *into* in place of *from* in 45 and *to* in place of *from* in 46) results in absurd literal meanings (it also needs to be noted (*20*) that the perception of the absurdity of a statement is much weaker through a language different from the mother tongue). Case 47 pertains to the solution procedure of a chemical thermodynamics problem, and the use of *at* in place of *from* affects the subsequent calculations, as the student does not consider a temperature increase, but a process at the constant 298 K temperature. The very frequent use of the preposition *in* (sometimes *under*) with reference to models (48, 49) relates to oversimplifications in the mode of expression and simultaneously to poor familiarity with the bases of the scientific method – in this specific instance, with the very concept of *model*.

44. The work is done by the system *of* the surroundings.
45. This experiment is based on heating the ice to change the ice *into* solid state to liquid state and gas state.
46. The heat are need to convert ice *to* solid to melting point and boiling point.
47. The water is heated *at* 298 K to 373 K.
48. *In* real gases many equations were proposed.
49. Gases *in* real gases behave differently.

Difficulties at Expressing Comparisons

Comparisons are fundamental for all of the components of a chemistry course, from the interpretation of experimental results in laboratory practice to the most theoretical discourses. The way in which comparisons are expressed is different in different languages. Most UNIVEN students find the expression of comparisons through English particularly difficult, up to the point that many statements would suggest a practical absence of the very concept of comparison. Although this is not the case (at least, as long as everyday life is concerned), it appears that having come into contact with the use of comparisons in chemistry (and in the sciences) only through a language that they do not master has prevented the development of the perception of the role of comparisons within the scientific approach. This complicates not only the direct expression of comparisons between quantities, but also the understanding of concepts that imply comparisons between values, like the relative values of certain physical quantities or the relative tendencies to do something. In some cases, the error can be viewed as solely grammar-related and it is possible to assume that the student has a basic understanding of the concept; e.g., the adjective does not have the *er* ending in cases 50 and 51 and is completely missing in cases 52 and 53:

50. Copper has a *small* standard potential than silver.
51. Copper has a *high* tendency than silver to be in an oxidized state.
52. Redox reactions that occur spontaneously is due to some elements' tendency to be in an oxidized state than the other.
53. Test number 5: when zinc metal was dipped in a solution of copper sulphate, the reaction was fast because zinc has the tendency to be in an oxidized state than copper.

In other cases, the awareness that the concept in question implies comparisons appears to be totally absent. This may result in misunderstanding of what needs to be done at treatment-of-results level in a laboratory report. E.g., students who write

54. Objective: to determine the tendency of an element to be in an oxidized state

for the experiment in which selected metals are dipped into solutions containing ions of another metal, often fail to compare the pairs of metals involved in each test (the solid metal dipped into the solution and the metal whose ions are dissolved in the solution), thus missing the major goal of the experiment. This may also be related to the frequent neglect (or inadequate awareness) that a comparison implies at least two terms. E.g., case 55 does not relate the *higher tendency* of the metals used in the listed tests to the metal whose ions are present in each of the solutions, and the lack of this comparison unavoidably results in the students' failure to make a concluding comparison of all the metals utilized (listing them in order of decreasing or increasing relative tendencies to be in an oxidized state). Statement 56 (which followed the list of the chemical equations for the tests in which a reaction was observed) does not make any comparison, showing unawareness that the *relative tendency* concept implies a comparison by its very nature. In case 57, the absence of comparison appears to imply that only powerful reducing agents can react, showing unawareness of the fact that the reaction is driven by the existence of a *difference in the relative tendencies* to be in an oxidized state, not by one species having a *powerful tendency*). The unawareness that *relative tendency* implies comparison is even more sharply highlighted by case 58, where the *relative* concept is totally missing, leading to incorrect chemistry.

55. The metals used in test 1, 2, 10, 11 and 13 they have *higher tendency* to be in an oxidized state.
56. Based on the observations, the elements of chemical reactions that are mentioned above have a *relative tendency* to be in an oxidized state.
57. When two elements of different tendencies come into contact the result is that *a powerful* reducing agent will dissolve into solution.
58. The element have tendency to be in an oxidized state if its oxidation number is high.

Inadequate perception/internalization of the fact that comparisons require at least two items (and, therefore, a plural) surfaces in many sentences and is often coupled with the frequent neglect of the conceptual and grammatical distinction between singular and plural forms:

59. The compressibility in the real gas its depends on temperature and pressure. The dependence is different for different *gas*.

In other cases, the *comparison* term is utilized with reference to different kinds of analysis, like the study of dependencies, showing inadequate awareness of the fact that only entities having the same nature can be compared:

60. The heating curve is the curve that is determined by *comparing* the temperature and the time.

The difficulties with the comparison concept heavily affect the answers to questions asking students to compare two or more cases. An illustration is offered by the answers to a question considering an isothermal expansion and an adiabatic expansion of the same gas starting from the same conditions and reaching the same final volume (respectively sub-questions b and c of a multi-step question), and asking first to calculate the final pressure, the work done and, for the adiabatic case, also the final temperature, and then to compare and discuss the results obtained in the two cases: 34 students out of 62 did not answer the sub-question requesting comparison, and most of the others did not compare the values obtained from the calculations, but tried to reproduce some memorized bits concerning the two types of processes, or only one of them (61, 62), or wrote totally random statements (63, 64):

61. Since the process is reversible and adiabatic, then no exchange of heat is possible, i.e., dq = 0.
62. For the two processes the conditions should be calibrated so as to suit the condition of process to take place accurately.
63. For a reversible process, work is done by the surrounding on the system. For a reversible adiabatic process, work is done by the system on the surrounding. For process b – work is done by the surrounding on the system. For process c – work is done by the system on the surrounding.
64. The gas in b expands reversibly and in c it expands reversibly and adiabatically. In c exothermic reaction is taking place. Work is done on the system by the surrounding. In b is endothermic reaction. The system absorbs heat from the surroundings. Work is done by the system.

The problem persists even at advanced level, as clearly highlighted, e.g., by the total absence of comparisons in most answers to the question "Compare the energy levels of a particle in a one dimension box and the energy levels of a harmonic oscillator (comparing means that you discuss both the similarities and the differences)", where the specification of the meaning of *comparing* was meant to help students; the answers were restricted to the reproduction of some memorized bits:

65. Schrödinger equation for a particle in one dimension: $-h^2/2m\ (d^2\psi/dx) = E\ \psi(x)$; $k\ (emE/h^2)^{1/2}$.
Solutions of the Schrödinger equation: $\psi = A\ e^{ikx} + B\ e^{-ikx}$.
66. For harmonic oscillator, $E_v = (v + ½)\ \hbar\ \omega$
$\Longrightarrow E_v + E_{v+1}$ levels $\Delta E = (v + 3/2)\ \hbar\ \omega - (v + ½)\ \hbar\ \omega = \hbar\ \omega$
For E_v, the zero point energy $E_o = ½\ \hbar\ \omega$.

Outcomes of the Simultaneous Presence of Different Errors

Among the examples considered so far, cases 1−62 illustrate language-related difficulties in the selection of words, or of functions like prepositions, through the consideration of simple (often one-clause) sentences. The last four cases (63−66) clearly highlight the fast increase of expression difficulties and errors as the logical complexity of the expected answer increases. Actually, the simple "sum" of few (often just two) of the errors considered in cases 1−62 suffices to yield statements in which the chemistry gets more and more confused or lost, and the literal meaning gets less and less traceable, up to often becoming absurd. Providing even a barely representative sampling of possible combinations would require much more than the available space and the illustrative role of such examples would be less focused, as they would basically constitute examples of statements and descriptions in which both the chemistry and the literal meaning would become too confused to be distinguishable. At this stage of the chapter, it is therefore more informative to focus on *how language-related difficulties hamper the development of skills that are essential for chemistry understanding*

– an investigation based simultaneously on the analysis of students' writings and on the information from classroom interactions.

Language and the Acquisition of Relevant Skills

Information from Classroom Interactions

Classroom interactions are fundamental to diagnose students' difficulties in an immediate direct way: not mediated through the attempts at interpreting what a student has written and making hypotheses, but attained through talking and, while talking, exploring details and possibilities in all the directions that appear significant. For the interactions to exist and be beneficial for the students and informative for the teacher in a situation in which students mostly refrain from talking in the class because the shyness associated with the awareness of not being in a position to make proper English sentences (sentences sufficiently correct to convey a meaning) blocks communication from their side, it is necessary to continuously invent approaches to simultaneously promote interactions and respond to the difficulties detected at each moment for the issue under consideration at that given moment. Intended as real-time response to the details surfacing through interactions, such design is unavoidably *ex tempore* and, therefore, it is not documented day by day; but the sum of all the many pieces of information surfacing from interactions and students' responses yields a picture highlighting not only individual difficulties, but causes and patterns. These are matched with the outcomes of the analysis of students' writings, leading to mutual verification of the information from the two sources (classroom interactions and written works analysis). In this way, classroom interactions offer apt keys for the interpretation of the difficulties highlighted by students' writings, and the collective analysis of issues (concepts, alternative conceptions, errors (*31–33*)) carried out within interactions brings to light a number of details of the problems encountered by students, thus facilitating the design of interventions.

Besides verbal communication, the interactions often involve short questions that students answer in writing and that are analyzed immediately afterward. The option responds to the known significance of "writing chemistry to understand chemistry" (*34*, *35*) – a significance that by its very nature links to the concept of chemistry as a language – and to the reflection-stimulating role of in-class questions (*36*). The option also has other important pedagogical roles, as it engages all the students (including those who do not participate in verbal interactions) and the teacher has the opportunity to talk with each student about what he or she is trying to write. It stimulates students' awareness of their problems or weak points. It provides the teacher with precious information about what aspects need to be clarified, or what errors need to be corrected through discussion – a discussion that is not delayed in time, but comes immediately after students have written their answers, while they are still focused on the effort to find a suitable answer. It helps foster fundamental skills, like how to read a question and understand its actual meaning.

The information that will be discussed in the next subsections, about the roles of language in the acquisition of important skills, is largely based on classroom interactions or confirmed through them. Its organization into subsections is rather artificial, because of the extensive overlaps among the domains of different skills and the conspicuous mutual enhancements among difficulties related to different skills; it is, however, expedient to limelight the key features for each type of ability and the significance of language skills in their acquisition and development.

Familiarity with Logic and Logical Frameworks

Language and logic are in a relationship of tight mutual dependence: logic is expressed through language and the backbone of sentence construction relies on logic. Our way of thinking does not consider only isolated pieces of information, but identifies relationships among them. These relationships are expressed through languages. Different languages have developed different options for the expression of specific logical relationships. The *natural* process of familiarizing with logic would be learning about the nature and meaning of the different logical relationships through the mother tongue and then expanding to express them through other languages, by learning to identify the tools through which a given other language expresses them (*37*). When the study of the mother tongue is not sufficiently articulate or complete to comprise learning about logical relationships, the familiarization with them becomes extremely difficult (*38*). This unavoidably results in serious lack of communication, as students fail to perceive the nature of the relationships between different pieces of information. At pre-university level in second-language disadvantaged contexts, the inadequacies often concern both the teacher and the student, because of the severe shortage of qualified science teachers and the additional difficulty that they encounter at expressing the content through English. The resulting scenario is effectively depicted in (*39*):

> *When teachers and learners cannot use language to make logical connections, to integrate and explain the relationships between isolated pieces of information, what is taught cannot be understood – and important concepts cannot be mastered.*

Having had several secondary school science teachers as my students at UNIVEN (within an important provincial government program to upgrade secondary school science-teaching by offering selected – supposedly the best – science teachers financial support to get a degree in the science that they are teaching) has offered the opportunity of direct contact with the extent of the problem from the teachers' side. Many teachers showed the presence and impact of language-related difficulties comparable to those of the younger students. The following statement, from a work by one of them, a secondary school chemistry teacher, suffices to offer a concrete illustration of the teacher's side of the scenario depicted by the previous quote:

67. In order for the reaction occurs the molecules of reactants must be more closer to each other and the collision between the reactant increases

because of concentration of reactants. The reaction will take place at a short time.
Rate law of concentration reaction: rate = k $[A]^p [B]^q [C]^r$
A, B, C are the reaction of reactants and p, q and r are the reaction orders respect of A, B, C.

Attempting to integrate the teaching of logical relationships within the teaching of course content at the tertiary level is not easy when the very concept of logical relationships is absent. A first-approach diagnostic question asking students to identify the correct statement between "I take the umbrella because it is raining" and "It is raining because I take the umbrella" yields disappointingly close proportions of students selecting one or the other statement, often with the majority selecting the latter. Students' use of logical connectives expressing cause-effect relationships (like *because, since, due to*, etc.) and of the verb *to cause*, are clear testimonials of the difficulties. The connectives are often utilized in a random way (68, 69) or the identification of the cause is incorrect (70–72; 71 is actually a sort of tautological expression, as the dependence on a quantum number is the meaning of quantization (the way it is expressed mathematically), not its cause; both 71 and 72 fail to consider that the cause of quantization is the presence of boundary conditions). Cases 73 and 74 are typical examples of tautological reasoning.

68. *Because* the molecules of water are too close, the heating curve of water make a slope.
69. The reaction has to reach the certain point *because* the transformation of reactant into products is successful.
70. Electrolytic cell also involve redox reactions only *because* spontaneous reaction is not possible *because* the reaction takes place in a single vessel.
71. Quantization arises *due to* the dependence of the solutions on the quantum number n, l, m_l.
72. When a particle is tunneling a potential barrier, its energy is decreasing, therefore it is not quantized.
73. The entropy change when two ideal gases mix is positive *because* when gases mix the entropy is always positive.
74. The rate of a chemical reaction decreases as the reaction proceeds, *because* reaction rate decreases with time as the reaction proceeds.

Other logical relationships, like hypothesis-thesis or condition-consequence (*38*), prove even more difficult to perceive, understand and express. The overall impact is poor understanding of the scientific approach as a whole – in particular, poor understanding of the meaning of the various models encountered, as models are based on hypotheses. The consideration of other important aspects of models (e.g., the validity range of a given model, or its limitations) too often remains beyond students' ability to follow the logic of a discourse. It is particularly sad when, on trying to expand a conceptual discourse during interactions with interested and potentially gifted students, they inform that they perceive how interesting the envisaged expansions could be, but are not able to follow them through English (*40*). All of this sadly restricts the objectives of courses such as physical chemistry to the pursuit of the attainment of basic physical chemistry literacy, without achieving real contact with the nature of physical chemistry reasoning and, therefore, without stimulating the type of interest that could prompt some students to consider it as a possible option for their future career (*26*). In this way, difficulties born from inadequate language-mastering prevent a real familiarization with one of the core branches of chemistry, and limit the possibility of training new physical chemists by limiting the number of possible candidates.

Visual Literacy and Communication through Imagery

Visualization is a form of communication. It is more immediate than communication expressed through language for aspects like the shape of objects, or details of their structures. Differently from language, it does not convey relationships among different pieces of information, unless it reaches sophistication levels that require specific professional training (like the drawings used in engineering or the flow-charts utilized in computer programming). It plays fundamental roles in science learning, as it provides external representations that facilitate understanding (*41*) and foster the generation of mental images – essential components of human mental processes in general and scientific elaboration in particular, with important functions in the learning process (*42*).

Two visualization domains are fundamental in tertiary level chemistry: the visualization of the invisible entities of the microscopic world of molecules to promote familiarization with their nature and structures, as well as a concreteness perception; and the use of diagrams to visualize and analyze trends from experimental data or to express mathematical relationships. The former responds to the simpler concept of using images of objects to draw attention to their characteristics, while the latter requires additional skills, interfacing or integrating with mathematical thinking.

The observations carried out at UNIVEN clearly highlight the interdependence between language-mastering and visual literacy (*43*). Most incoming students have poor (or very poor) visual literacy, and attempts to develop it bounce against the obstacles posed by poor language mastering. Interactions suggest that poor language-mastering is a major cause behind the poor development of visual literacy at pre-university level. Showing how to read an image (e.g., ball-and-stick models of molecules) requires the possibility of discussing the concepts related to that image, to highlight how each conceptual detail is represented by a corresponding detail in the image or, conversely, how each detail of the image has a meaning; therefore, it requires adequate mastering of the language through which explanations are given or the discussion is conducted. Inadequacies in the two abilities are mutually enhancing: language-mastering is needed to learn to understand how an image conveys information, i.e., to acquire the bases of visual literacy; visual literacy is needed to grab the information conveyed by an image and language-mastering is needed to express it, or to reflect on it (as our thoughts develop through sentences, i.e., through language). Similarly, for a student to be able to draw his or her own images to represent something, language-mastering is needed to acquire basic knowledge about the given entity, i.e., to attain sufficiently clear ideas about the information that is to be communicated by the image, and visual literacy is needed to identify the pieces of information that can be conveyed through the image (including details) and to be able to actually convey them.

The previously mentioned short questions asking for written answers during classroom interactions also include the drawing of images (when relevant to the theme considered at a given moment), in view of the importance of images as explanation tools and tools for classroom interactions, and of the significance of including image-related errors into error analysis options (*44*), and also of the valuable feedback on students' difficulties that they can provide (*45*). The mere drawing of the structures of rather simple molecules using ball-and-stick models highlights the presence of considerable difficulties in relating the information expressed through language and the information expressed through an image (relating the information expressed through words to the details of the image that students are drawing). If the tendency to follow the formula as literal guidance (for which students draw and link atoms in the order in which they appear in the formula (*12*)) can at least partially be ascribed to inadequacies in abstraction capabilities (as the symbolism of formulas is inherently abstract), drawing atoms with more bonds than their usual bond-formation ability (e.g., two or three bonds for a sphere representing an H atom) is an actual index of visual literacy inadequacies, because the mere counting of sticks should prevent errors of this type if the skill to relate the details of an image to some physical meaning (e.g., a drawn stick to a chemical bond) had developed sufficiently. Although in this specific case, the student's attention can be easily stimulated by asking how many bonds a given atom can form and how many appear in his/her drawing, the frequency with which these errors appear (even at 2nd and 3rd year level) is a revealing symptom of the impact of not having acquired a basic habit to relate conceptual information to the details of an image and vice versa.

Students usually need guidance to go beyond the tendency to draw and link atoms in the order in which they appear in the formula. This guidance is provided through information about the key features of the molecular structure, worded in such a way as to state general-type information and to stimulate/require reflection. This usually corresponds to a list of features, each of them expressed through a one-sentence clause; for instance, for the molecules of oxygen-containing acids, the clauses inform that the non-metal atom is central in the molecule, that the O atoms are bonded to the non-metal atom, that no H atom is bonded to the non-metal atom and that no O atom is bonded to another O atom (*46*, *47*). Translating this information into the details of an image requires simultaneous consideration of all of the pieces of information; this appears to pose demands analogous to those posed by complex sentences, and students usually require additional guidance to understand the structural implications of the individual pieces of information and of their combinations.

The ability to associate mathematical relationships to the diagrams representing them, and vice versa, depends on language-mastering, visual literacy and abstraction abilities simultaneously. The optimal basis would correspond to the gradual building of a mental term-image association for each of the most common relationships (linear, parabolic, exponential, hyperbolic, etc.), as this would help overcome abstractness perceptions in relation to these terms and concepts and, simultaneously, ensure a skill that is fundamental for analysis and interpretation. The frequency with which students interchange key terms (e.g., utilizing *exponential* or *hyperbolic* for any non-linear diagram, or calling *direct proportionality* any increasing diagram and *inverse proportionality* any decreasing diagram, including linear ones) testifies to the absence of such mental associations. Although some of these confusions may – at least partially – have the same roots as other confusions between terms, like the ones discussed in previous sections (e.g., *parabolic* and *hyperbolic* might be perceived as close to homophone), the absence of term-image mental associations can be considered the key factor. These are technical terms and, therefore, they are learnt as new terms (terms introduced in the class) in any instruction contexts, including mother tongue ones; on the other hand, the acquisition of the corresponding term-image mental associations depends on there being a specific focus at the teaching and learning level, and on the students' ability to understand explanations – an ability that is largely conditioned by their language-mastering level. The insufficiency or absence of mental images for key mathematical dependences seriously affects the student's ability to analyze data or discuss trends.

Not having acquired the habit and skills to relate information through language and information through graphical representations makes it difficult for students to identify the physical meaning of the details of the diagrams utilized in chemistry. The interpretation of the diagram of the isotherms of a gas offers clear illustrations of a variety of frequently encountered difficulties: cases 18 and 25–27 offer straightforward examples of the difficulties of reading the diagram (case 27, even speaking of *melting*, that has nothing to do with the diagram); case 75 highlights the failure to identify the correspondence between a horizontal segment and the constancy of the quantity reported on the y-axis; case 76 highlights the absence of clear ideas about what can be shown by a diagram and what cannot, or what is not shown by a specific diagram – a problem that, in combination with language-related difficulties, often leads to meaningless statements (statements not having an identifiable meaning):

75. Below critical point *the pressure increases* as there is a liquefaction which is the horizontal line.
76. Intermolecular interactions affect in the graph when they dissect.

At the drawing level, the most frequent error is that of attaching all the horizontal segments to a single curve on the right part of the diagram, which highlights a failure to understand the very nature of the diagram – the representation of several curves, each corresponding to a specific temperature value and which, therefore, cannot merge into one. Explaining the nature of this error is particularly demanding because, even in its most simplified wording, the communication requires a level of language-mastering that is beyond the one possessed by most students at UNIVEN.

Other frequently observed difficulties in relation to the drawing of diagrams concern the identification of the cases where a trend is asymptotic and the cases where the graph encounters one of the axes (e.g., that the diagram of the concentration of a reactant versus time as a chemical reaction proceeds is not asymptotic to the y-axis, but starts at a specific value, corresponding to the initial concentration and to time zero). Difficulties of this type are also closely related to inadequate perception of the physical meaning of the details of a diagram – an ability that only appropriate training through explanations (i.e., through language) can develop.

When both the conceptual demands of the course material and the imagery sophistication increase, as in the process technology course, the difficulties related to the interplay between communication through language and communication through imagery increase. Translating the description of a process into a flowsheet, or a flowsheet into the description of a process (and, even earlier, into the understanding of the nature of the process) proves difficult for most students.

The Development of Abstract Thinking Abilities

Abstract thinking is required in all of the physical sciences. It may involve aspects like generalizations based on observed phenomena or trends, or the use of identified relationships between different pieces of information (e.g., cause-effect relationships) to propose hypotheses, or the large domain of mathematical descriptions. Abstract thinking is essentially language-constructed and, therefore, it requires adequate language-mastering sophistication.

Language-related difficulties toward abstraction are highlighted in students' works not only by the meaning of statements, but also by more basic features like the random, nearly always incorrect use of abstract terms. It is the case, e.g., of the frequent absence of distinction between an adjective (or an adverb) expressing a quality and the associated noun denoting that quality, like *different* and *difference* (77) or *spontaneous* and *spontaneity* (78), or case 79, where *massively infinity* replaces the *infinitely massive* present in the textbook:

77. The diagrams are *difference* due to the amount of ice measured.
78. The reactions in tests 4, 6, 8, 9 and 12 are *spontaneity*.
79. In a reduced mass technique or system, the nucleus is considered as *massively infinity* (M) therefore it fixed.

Although the confusion between terms perceived as homophones may be at least partially responsible for such interchanges, classroom interactions highlight a diffuse absence of the perception of the distinction between the two categories of terms and concepts and the corresponding roles. Similarly, difficulties in distinguishing between information that relates to a physical object and information that relates to a property (denoted by an abstract noun) appear evident in a number of circumstances, and are highlighted in the sharpest way by error-analysis exercises (*31–33*). E.g., on being asked to detect and discuss the error in the statement "In an electrolytic solution, the total number of positive ions is equal to the total number of negative ions" (the error being the use of *ions* in place of *charges*), most students considered that the error was the reference to an electrolytic solution, and that the correct reference would be *electroneutrality*:

80. Not "in an electrolytic solution", but "in electroneutrality". In electroneutrality the total number of positive ions is equal to the total number of negative ions.
81. In an electroneutrality of a chemical system it is electrically neutral, and therefore the total number of positive ions is equal to the total number of negative ions.

The diffuse habit of using the preposition *in* with reference to models (48, 49) may contribute to decrease the perception of the distinction between objects and abstract concepts or models, which underlines how basic grammar incorrectness may affect many levels of understanding and conceptualization.

Diagrams can be viewed as an overlap between visualization and abstract thinking, as they do not represent objects, but constitute the visualization of trends, of dependence-types and of other features pertaining to our modeling activities. An extreme example of the difficulties at perceiving the abstract nature of diagrams is reported in (*29*): a student writing that the potential well is *a sort of container*, because of interpreting the diagram as representing a physical (every day life) object. The difficulties toward the handling and interpretation of diagrams, outlined in the previous section, can simultaneously be ascribed to visual-literacy inadequacies and to abstract-thinking inadequacies. The difficulties at communicating information about the meaning of a diagram or the analysis of its details relate both to general language-mastering shortcomings and to shortcomings regarding the language-mastering sophistication-level that is needed for abstract thinking.

Abstract thinking depends on the mastering of other abilities, first of all the ability of following (identifying and/or building) logical frameworks. Therefore, it requires a rather sophisticated use of language, capable of understanding and utilizing complex sentences, i.e., sentences consisting of more than one clause and expressing logical relationships among different pieces of information through the logical connectives between clauses. Promoting and developing abstract thinking abilities would therefore first of all require overcoming the difficulties toward the expression/communication of logical relationships (discussed in a previous section). This factor alone provides unequivocal evidence of the importance of mother tongue instruction to foster the development of abstract thinking abilities since pre-university level.

The Familiarization with the Scientific Method

Adding to the analysis of students' writings from the integrated points of view of language aspects and chemistry aspects (*21–24*) is a parallel analysis conducted with specific focus on method-related aspects having key roles for basic chemistry understanding: the distinction between general and particular

(*48*), the distinction between systems and processes (*49*), the distinction (*50*) between physical quantities and their changes (e.g., *S* and ΔS, *H* and ΔH, etc), the distinction between values and numbers (*51*), the understanding and expression of the cause-effect relationship (*52*), and the distinction between the microscopic and the macroscopic descriptions in chemistry (*53*, *54*). The analysis also included phenomena not falling within expected method-related categories, like the dominance of the *reaction* concept and the associated terms, replacing a number of other concepts, and corresponding terms, in a variety of occasions (*55*). This analysis showed that language-related difficulties play major roles in the observed confusions and in the students' failure to acquire an adequate perception of the distinction among different categories of entities, domains or tools and the corresponding roles. A brief overview of the major aspects investigated may serve to highlight the major links between observed problems and language-mastering shortcomings. The failure to distinguish between what has general validity and what refers to individual (particular) cases is rooted in the failure to perceive the way in which the language of instruction expresses this distinction – a failure preventing the very development of the awareness that such distinction exists. The use of verbs or qualifications typical of processes in relation to objects (e.g., cases 21, 22, 38, 39), or vice versa, has clear language roots: it would not occur in the mother tongue because the perception of what can be referred to an object and what cannot pertains to the ensemble of concept-language correspondences that are deeply internalized in the mother tongue. The random use of the names and symbols of physical quantities and their changes largely relates to the tendency toward wording oversimplification by omitting key words (in this case, the word *change*) and, to some extent, also to the difficulties with comparisons (because a change implies a difference); but it results in confusions that may seriously affect both conceptual understanding and the problem solving stage (above all in the chemical thermodynamics course). The confusion between *numbers* and *values*, for which students tend to call any numerical value the *number* of something, without relating it to its physical meaning in the given context and/or to its dimensions, also affects the problem solving stage; e.g., when students call the mass of a given sample *mass number* (and write so in their data list), they often do not treat the value as a mass value in the problem solution procedure, and their discussions may show confusion between the *mass of a sample* concept and the *mass number* concept. The distinction between the microscopic and the macroscopic levels of description appears to be arduous for chemistry students worldwide, but language-related difficulties bring additional complications; e.g., inadequate perception of word coupling correctness increases the rate at which macroscopic properties or phenomena are ascribed to microscopic entities (82) or vice versa:

82. Water molecules consist of three phases which are reached in different temperature.

The tendency to use the terms *reaction, react, reactant*, etc. for any type of process, or in any other occasions in which students are not sure of the correct term, relates to lexicon learning difficulties, but also to insufficient awareness of the distinction between *general* and *particular*, specifically missing the general character of the *process* concept (83, 84) or transferring (85) to the general *system* concept the frequent misconceptions on chemical equilibrium (i.e., on chemical reactions):

83. A *process* is a *reaction*.
84. A spontaneous *process* is a *reaction* that occurs without the intervention of our technology.
85. A *system* is at equilibrium if the concentrations of all the species are the same.

In summary, language-related difficulties hamper students' perception of fundamental aspects of the scientific method, of the significance that they have for investigation, descriptions, discussion and modeling in the sciences. On one hand, specific familiarization with these aspects needs to initiate at primary and secondary school levels, for students to attain at least the perceptions and internalization corresponding to basic literacy. On the other hand, when in-class communication breaks down because of the limitations posed by inadequate mastering of the language through which the communication is expected to be realized (*39*), opportunities for familiarization with the basic aspects of the scientific method simply do not exist.

De Facto Annihilation of the Potentialities of the Tools That Could Help Lower the Impact of Language-Related Difficulties

The previous subsections have considered the skills that are important for chemistry understanding and for the acquisition of chemical knowledge individually, to better highlight how the acquisition of each skill depends on language-mastering. The way in which different skills contribute to the understanding and knowledge-acquisition process is cooperative: each skill contributes an understating pathway that supports and reinforces the other pathways. Because of this, inadequacies in one of the skills may weaken the efficacy of the others. When the common background root of inadequacies is inadequate language-mastering, skills mastering may remain below an efficacy minimum threshold and, therefore, the potential benefits of the whole ensemble of skills may fade away. Attempts to develop them within tertiary level courses are hampered by the lack of prior exposure and by the persistence of language-related difficulties.

The mutual enhancement of inadequacies generates a sort of downward loop leading to a steady decrease in the quality and extent of mastering skills. For instance, up to a few years ago it was possible to obtain some benefits from in-class collaborative constructions of diagrams (flow charts) to represent logical frameworks that students found too difficult to identify and follow by reading. Reading the text in a stepwise way and adding a new box to the flowchart in correspondence to a new piece of information had the double role of highlighting the logical framework concerned and of fostering some text-analysis abilities

(*56*). Recent attempts of this type encountered complete failure due to the students' inability to understand the very meaning of relating different pieces of information, or to perceive the representation offered by the diagram as a form of communication highlighting these relationships. The idea of a logical framework (a set of relationships between different pieces of information, building an overall picture) appeared to be totally alien, showing the ultimate (and extreme) consequences of the absence of exposure to logical relationships; verbal explanations aimed at stimulating basic awareness about the existence or possibility of such relationships bounced against inadequate language-mastering and failed to bring benefits. A recent attempt with a third-year group, focusing on the reasoning that led chemists to conclude that electrolytic solutions contain ions, can illustrate the situation in a concrete way. The reasoning (that students find difficult to understand from a text, even after the wording has been simplified and its logic itemized) can be represented by a simple diagram (*12*) singling out the two sets of experimental information and related inferences (the ability of these solutions to conduct electric current, prompting the inference that they must contain dissolved charged particles, and the fact that the magnitude of their colligative properties is greater than what would be expected on the basis of their concentration, prompting the inference that the number of particles dissolved is greater than what would be predicted on the basis of concentration) and the overall inference that the solute dissociates into charged particles (ions). The construction of the diagram was totally guided, as students did not offer any suggestions because they did not understand the meaning of the text; each relationship was discussed in detail, both because of the need to clarify all the aspects involved and because the whole issue underlines – in a concrete way – the role of experimental information, and of the inferences derived from it, in the building and progress of chemical knowledge. On subsequent testing at the classroom interaction level in the next lecture, it turned out that students had memorized the text and had memorized the diagram, but were unable to recognize or discuss any correspondence between the text and the diagram. Although the outcome may be counted as just another of the countless cases of passive memorization and passive regurgitation, one needs to recall that generalized passive memorization originated first of all as a response to failure to understand the literal meaning of texts, i.e., it is dominantly caused by language-related difficulties.

The failure of the experiments just described concretely illustrates the interdependence of the various skills, and of the benefits that can derive from each of them. Visualization can help understand and clarify the information transmitted by a text on the condition that the student is able to interpret it, i.e., to perceive it as a communication form and to catch the information it conveys – a type of ability whose development depends on language-mastering. Suitable use of visualization can stimulate and facilitate the development of logical thinking and abstract thinking (*57*) abilities, but, again, on condition that the student is able to perceive the messages it conveys. In summary, the common denominator of the identified problems is inadequate language-mastering, hampering the development of those abilities that could help lower its impact. When the inadequacies in all of these abilities are serious, the unavoidable – and sad – consequence is a factual limitation of the presentation of the course content to a set of discrete pieces of information, without the possibility of linking them into articulated conceptual pictures. The ensuing losses in terms of students' approach to chemistry and its methods are self-evident: as too many connections between components remain beyond their reach, they are deprived of a real contact with the encompassing and multifaceted nature of chemistry.

Approaching Chemistry through a Second Language in Non-Disadvantaged Contexts, at Pre-University Level: Inherent Constraints and Their Impacts

The familiarity with the situation described in the previous sections, and the ensuing awareness of the paramount importance of language-related aspects in the teaching and learning of chemistry, unavoidably generate concerns about the risk that the same difficulties might extend to other contexts through unpredicted (although not unpredictable) consequences of specific choices. Such concerns are particularly serious in relation to the realization of attempts to teach chemistry (and other sciences) through languages different from the mother tongue, in contexts where mother tongue instruction is the general option. A clear illustration of the motivations for concern is provided by the descriptions regarding wording and language in the outline of attempts of this type in Italian secondary schools, presented at an Italian Chemical Society Conference on Chemical Education (*Chemical Thought and Scientific Education in School Reforms*, Assisi, 9–11 December 2004) by secondary school teachers involved in such attempts. The teachers explained that introducing chemistry through English, instead of Italian, required the use of very simple, one-clause sentences like "The atom consists of a nucleus and electrons. The electrons move around the nucleus." and the like. Although the presenters appeared to consider this as positive, the reasons for concern are immediately evident in the light of the observations from disadvantaged contexts described in the previous sections. The choice of using a foreign language that pupils do not master beyond a certain level forces the teacher to turn chemistry into an ensemble of discrete, simplified pieces of information, not linked to each other by logical connections. This implies: (a) the loss of logical relationships and logical frameworks and, therefore, the impossibility of presenting chemistry as an articulated discourse; (b), the impossibility of highlighting «observations → inferences → models» sequences and, therefore, the impossibility of discussing the generation of models; (c) the impossibility of highlighting method-related aspects and, therefore, the impossibility of highlighting the investigation approaches typical of chemistry. In other words, it implies the impossibility of a real exposure of pupils to the nature of chemistry.

It may be questioned whether, and to what extent, the oversimplification outlined in those presentations is unavoidable, but simple considerations show that it is. If explanations need to be adapted to the pupils' language-mastering level of a given foreign language, and since this level is unavoidably more limited than for the mother tongue, oversimplifications of this type are bound to become an unavoidable feature (or methodological constraint) of the use of a foreign language in the first exposure of pupils to chemistry (or to another science). Besides preventing pupils from real contact with the nature of chemistry (or of another science), such oversimplified presentations of chemistry fail to respond to the general objectives of Higher Secondary School education – the education to critical and creative thinking, which requires logical connections between different pieces of information as parts of its necessarily inherent components.

Learning the Language of Chemistry

Language Mastering and Science Learning

All of the information presented and discussed in the previous sections shows the fundamental importance of language-mastering for chemistry learning (and science learning in general). Attaining the needed level of language-mastering comprises two major routes: the acquisition of general language mastering and the acquisition of adequate mastering of the language aspects typical of scientific communication. The former involves the use of grammar, but also the ability to analyze sentences (an ability based on the recognition of the types of possible relationships between different pieces of information and the ways in which they are expressed within a given language, practically bordering with the foundations of linguistics); the acquisition of these abilities is pursued most effectively within the study of the mother tongue. The acquisition of adequate mastering of the language aspects typical of scientific communication implies the ability to utilize the selection of individual words, the combination of words into clauses and the organization of clauses into complex sentences as the major tools to pursue the fundamentals of scientific communication – the requirement of being rigorous and the requirement of being clear (*2*, *3*)). Fostering and building this ability pertains to the chemistry/science courses (ideally with cross-disciplinary cooperation (*58*) with the courses in study of the mother tongue).

Chemistry is an ideal area for the education to the *language of science* and, conversely, language aspects are particularly important in learning and understanding chemistry. Highlighting language aspects – utilizing them to clarify concepts – is a way of facilitating understanding and reflection: it is the concept of the science teacher as a language teacher, to foster the development of creative scientific thinking (*1*). Reflections on language aspects that can be key to enhancing understanding can be stimulated also through *ad hoc* questions in the chemistry textbook (*59*).

Cross-Language Character of the *Language of Science*

The *language of science* has an acknowledged cross-language character. Because of this, scientists can often read scientific texts in languages that they would not be able to utilize for conversation or for reading other types of literature. A scientist's ability to do so derives from two factors: the characteristics of language-usage in scientific communication and the scientist's general language-mastering level (with important contributions also from the use of symbols, for those sciences that make extensive use of symbol systems). The use of language in science communication has features that relate to the specific requirements of such communication (being rigorous, being clear) and to the fact that the characteristics of the objects of interest are language-independent. The criteria for selecting words and combining them refer to the characteristics of the object of interest and, therefore, are cross-language (language-independent) in character, although the individual words or the individual connectives are different in different languages. For instance, the description of a given molecule or a given chemical reaction, or the explanation of why the rate of a chemical reaction depends on temperature, will be basically the same in any language – not because of being literal translations of each other, but because they need to provide the same pieces of information. Being able (as a scientist is) to anticipate the nature of the types of information that one is searching for facilitates the possibility of comparing descriptions expressed through different languages, or of understanding a new description expressed through another language.

The extent to which a scientist develops the ability to read science in different languages largely depends on his or her general language-mastering ability, as the bases of language knowledge (the bases of linguistics) are the roots enabling the identification of meanings and correspondences in descriptions expressed through different languages, at least, as long as the identification is not complicated by features like the use of a different and unfamiliar alphabet, or a language pertaining to a totally different and unfamiliar linguistic group. In this way, the level of acquired language-mastering is a key also to the possibility of reading science texts in different languages.

Approaching Chemistry through the Mother Tongue

The mother tongue corresponds to the deepest internalization of features like the sound-concept correspondence, the meaning of words and the immediate identification of the meaning conveyed by word combinations in individual clauses and complex sentences. It corresponds to a sort of *mental home* (*60*) where understanding is easier, more immediate and more complete.

Approaching a science for the first time is not easy for a pupil. It implies a new way of approaching things and thinking about them; a new way of observing objects or phenomena, of posing questions about them and of searching for answers; a new type of information that becomes important and needs to be understood; a new set of tools to utilize for problem solving; and a new way of thinking in order to understand the interpretations and models that have been proposed.

Chemistry is considered particularly difficult by many pupils worldwide. Although inadequacies in the educational approaches often have significant weights in the generation of such perception, the complexity of chemistry as a science is a factor whose weight needs not to be overlooked. The *new ways* with which a pupil needs to become familiar upon approaching chemistry for the first time are numerous and demanding in terms of mental engagement and effort: the pupil has to start thinking in terms of composition and transformations, amounts of substances and proportions between them, structures and behavior of entities of an invisible microscopic world, symbolic representations and their correspondence to entities or to amounts. The addition of the mental effort inherent in trying to catch the meaning of sentences and explanations through a language that does not convey to him or her an immediate perception of meaning complicates the approach to chemistry enormously and may prevent the attainment of real contact. Utilizing the mother tongue means maximizing the possibility of understanding and the comfort with which it is pursued and attained. The observations discussed in the previous sections prove it as a *conditio sine qua non* for the pupil to be in a position to attain a satisfactory level of understanding and, simultaneously, to acquire those skills (logical thinking, visual literacy, abstract thinking) that are essential to understanding and that develop more effectively and completely through the mother tongue.

There are other general type reflections in favor of approaching chemistry through the mother tongue, which will be outlined only briefly here. Any science is a human endeavor; when a science is not expressed through the mother tongue, it becomes an endeavor pertaining to another community, not one's own (*3*); this has impacts on the perceptions that young people develop about that science (or about science in general, if all the sciences are approached through languages different from the mother tongue) – perceptions that will unavoidably diverge from the generally assumed universal character of science. Diversity is a fundamental source of richness: missing or losing the possibility of expressing science through certain languages (what is tantamount to missing or losing the possibility of generating science through those languages) would mean losing the possibility of benefiting from certain perspectives: it would mean an overall impoverishment for the future of science. These reflections, and the evidence collected from chemistry courses, thoroughly supports the importance of developing languages (like the African languages) so that they can express science fully (*3*); this will contribute to development in the continent (*14–17*) by raising general science literacy and the level of knowledge acquisition by students opting for science degrees; in the long run, it can be expected to contribute new perspectives to science through the richness of diversity. The same evidence supports the importance of maintaining mother tongue approach to chemistry (and to the other sciences) in contexts that are currently using mother tongue instruction.

Expressing Chemistry through Other Languages, at Pre-University Level

Globalization implies increased contacts across countries and cultures. Science has always had an across-border character with exchanges across any type of borders (national, cultural, religious, etc.). Exchanges require communication through a language that is understood by both parties. That is why, in the European history, science has always had a *lingua franca*: Latin for many centuries, then French, German, and currently English, that has become the *lingua franca* worldwide.

The familiarization with the *lingua franca* is very important and ideally should be pursued since early approaches to chemistry (or to other sciences). It is however necessary to pursue it without affecting the pupil's approach to chemistry, without making its understanding more difficult or limiting the understanding extent and depth. The key to simultaneously take into account the need to learn and understand through the mother tongue and the importance of being able to communicate through the *lingua franca* (and also through other languages) is the separation of the two major components of school learning: understanding, and the expression of what the pupil has understood. Understanding is enormously more effective through the mother tongue. Once understanding is achieved, the pupil can learn to communicate the just acquired knowledge through other languages, and the efforts to do so will imply additional reflections on the pieces of knowledge that he or she has to communicate, because expressing them through a different set of tools (those of a different language) requires an additional type of analysis. For instance, the pupil will not just learn that "The atom consists of a nucleus and electrons. The electrons move around the nucleus" and the like; he or she will learn about the models proposed for the atom and how and why the modern model has been attained; then, by trying to express this knowledge through another language, the pupil will have to try and word things in a simpler way than through the mother tongue, but without oversimplifications that abolish relationships between different pieces of information; then, the pupil will have to identify the nature of the relevant relationships to be able to express them through the other language. In this way, learning to express chemistry through another language will simultaneously contribute to clearer chemistry understanding and to the development of adequate mastering of the *lingua franca* or of other languages. Chemistry learning will benefit from the language-related analysis, and the general communication abilities of the pupil will expand.

Learning Chemistry through Other Languages, when the Student Is Ready

The ground prepared by the option just outlined will gradually put the student in a position to understand chemistry through another language directly, by setting adequate foundations for the recognition of language-chemistry correspondences. This is the experience of the many non-native English speakers who have used English textbooks at tertiary level (*61*), or the experience of students who have spent study periods in other countries. Once the student has acquired sufficient familiarity with the nature and approaches of chemistry and all the needed skills have been developed adequately, using a different language poses demands that stimulate reflections without interfering with the quality of understanding. Lots of items are the same in chemistry books (or chemistry explanations) whatever the language utilized. If the skills have been developed adequately, formulas and equations contribute to understanding, because the student is able to read them, i.e., to associate a meaning to them; diagrams and images contribute to understanding, because the student is able to interpret them; analyzing sentences in the other language to understand their literal meaning contributes to the understanding of the chemistry meaning, because of the intimate language-concept association; altogether, the different communication options (language, visual, symbolic) merge into a chemistry discourse in which chemistry as a language dominates over the peculiarities of individual languages. On this basis, learning more languages and using them to read or communicate chemistry enriches reflection and thinking perspectives.

Discussion and Conclusions

The evidence from second-language disadvantaged contexts shows how inadequate mastering of the language that is the medium of instruction hampers the acquisition of chemical knowledge and prevents a real contact with the nature of chemistry as a science. In this way, it also highlights the importance of the mother tongue for the familiarization with the foundations of the chemistry discourse and for the acquisition of those skills that are essential to effectively pursue such familiarization. The degree of language mastering within the mother tongue provides the optimal medium to facilitate understanding of the concepts and approaches that are typical of chemistry. The features concerning logical relationships, logical frameworks and other aspects of the science discourse transcend individual languages, and the role of the mother tongue is to foster sufficient awareness and familiarization to enable a student to recognize them and their communication functions; this will also constitute the basis for the student to become able to express scientific discourse through other languages. Similarly, the use of imagery as a communication tool transcends individual languages, but the ability to relate conceptual information to the details of an image, and vice versa, develops through the analysis of these relationships, i.e., through language, making the mother tongue the optimal medium to foster the development.

The general inferences – linking to the educational implications of the relationships between language and science – can be summarized as follows:

a) the importance of adequate general language-mastering for a student to be able to access the chemistry discourse, and the role of the theoretical knowledge of the mother tongue to attain the needed language-mastering levels;
b) the importance of utilizing the mother tongue to foster familiarity with the chemistry discourse and adequate mastering of skills like visual literacy, logical abilities and abstract thinking;
c) the importance of distinguishing between the understanding component of the learning process (to be realized through the mother tongue, at least until a solid basis is built) and the expression component, that can utilize any other language since the beginning, and;
d) the foundation character of the acquisition of language-mastering and the other relevant skills.

An ideal route to optimize both the acquisition of chemical knowledge and the acquisition of the ability to communicate in many languages (responding to the needs of increasing globalization) will utilize the mother tongue to facilitate understanding and to foster the development of the relevant skills, while simultaneously encouraging expression of the acquired knowledge through other languages. After the familiarity with chemistry discourse has reached a sufficiently solid foundation, the student will be in a position to benefit also from approaching new components or areas of chemistry through different languages because, then, he or she will know what to look for in a text expressed through a different language and how to identify pieces of information and the relationships between them, and the search from the different point of view provided by the different way of expressing things in the other language becomes an incentive to broaden the range of reflections. In summary, the mother tongue is necessary to the acquisition of adequate familiarity with the chemistry discourse and with the skills essential to it; once this familiarity is consolidated, adding the use of other languages may offer significant opportunities for mutual enhancement between language-mastering and the acquisition of chemical knowledge, up to a merging of the perspectives of the language of chemistry and chemistry as a language.

References

1. Sutton, C. *Enseñ. Cienc.* **2003**, *21* (1), 21–25.
2. Mammino, L. *Il Linguaggio e la Scienza*; SEI: Turin, Italy, 1995.
3. Mammino, L. *Terminology in Science and Technology*; Ditlou: Thohoyandou, South Africa, 2006.
4. Bradley, J. D.; Brand, M.; Gerrans, G. C. In *Widening the Scope of Chemistry*; IUPAC & Blackwell Scientific Publications: London, 1987.
5. Mammino, L. *ALDEQ* **1996**, *VIII* (2), 375–384.
6. Mammino, L. *Scuola Viva* **1997**, *8*, 14–17.
7. Mammino, L. *CHEMEDA* **1998**, *48–50*, 30–38.
8. Mammino, L. In *The Language of Science*; Seepe, S., Dowling, D., Eds.; Johannesburg, South Africa, 2000; pp 52–71.
9. Mammino, L. In *Edichem '99: Chemistry in the Perspective of the New Century*; Bari University: Bari, Italy, 1999; pp 212–230.
10. Mallinson, G. G.; Sturm, H. E.; Mallinson, L. M. *J. Chem. Ed.* **1952**, *29*, 629–630.
11. Lahore, A. A. *Enseñ. Cienc.* **1993**, *11* (1), 59–62.
12. Mammino, L. In *Visualization: Theory and Practice in Science Education*; Gilbert, J. K., Reiner, M., Nakhlekh, M., Eds.; Springer: Dordrecht, The Netherlands, 2008; pp 155–185.
13. Mammino, L. *Iter* **2001**, *10*, 44–48.
14. Prah, K. K. *Mother Tongue for Scientific and Technological Development in Africa*; German Foundation for International Development: Bonn, Germany, 1993.
15. Prah K. K. *African Languages for the Mass Education of Africans*; German Foundation for International Development: Bonn, Germany, 1995.
16. Prah, K. K. In *Between Distinction and Extinction*; Prah, K. K., Ed.; Witswatersrand University Press: Johannesburg, South Africa, 1998; pp 1–16.
17. Prah, K. K. In *Rehabilitating African Languages*; Prah, K. K., Ed.; CASAS: Cape Town, South Africa, 2002; pp 1–6.
18. Case, S. M. *Teach. Educ. New Countries* **1968**, *9* (1), 15–26.
19. Mammino, L. *Zimb. J. Educ. Res.* **1998**, *10* (3), 189–209.
20. Mammino, L. In *The Language of Science*; Seepe, S., Dowling, D., Eds.; Johannesburg, South Africa, 2000; pp 94–101.
21. Mammino, L. *J. Educ. Stud.* **2005**, *4* (1), 36–41.

22. Mammino, L. *J. Educ. Stud.* **2006,** *5* (2), 189–213.
23. Mammino, L. *J. Educ. Stud.* **2007**, *6* (2), 199–214.
24. Mammino, L. *J. Educ. Stud.* **2009**, *8* (4), 142–157.
25. In South Africa, the academic year starts somewhere during January and ends somewhere during December (summer being from 21st December to 21st March). The actual dates are different for different universities.
26. Mammino, L. In *Chemistry Education in the ICT Age*; Gupta-Bhowon, M., Jhaumeer-Laulloo, S., Li Kam Wah, H., Ramasami, P., Eds.; Springer: Dordrecht, The Netherlands, 2009; pp 197–223.
27. It may be that the sureness of stating "would never occur in the mother tongue" is also influenced by the fact that my mother tongue (Italian) is alphabetical, with strictly biunivocal sound-concept correspondence. But so are the local languages in Sub-Saharan Africa; therefore, errors of this type would not occur in the students' mother tongue. The mother tongue of nearly all the students attending UNIVEN is Tshivenda, the language of the Venda people and one of the 11 official languages in South Africa. The other official languages are Afrikaans, English, IsiNdebele, Isixhosa, Isizulu, Sepedi or Northern Sotho, Sesotho or Southern Sotho, Setswana, SiSwati and Xitsonga (the prefixes Isi, Se, Si, Tshi and Xi identify the "language" concept with respect to the main name, which is the name of the ethnical group; therefore, the languages listed after *English* are respectively the languages of the Ndebele, Xhosa, Zulu, Pedi, Sotho, Tswana, Swazi and Tsonga people). Although all these languages are official, instruction is carried out in English (except from some schools that use also Afrikaans, the language of the Afrikaners).
28. Clerk, D.; Rutherford, M. *Proceedings of the Sixth Annual Meeting of the SAARMSE*; UNISA: Pretoria, South Africa, 1998; pp 126–131.
29. Mammino, L. *Southern Africa J. Math. Sci. Educ.* **1995**, *2* (1,2), 69–88.
30. I do not speak any African language. For those experiments, I asked a non-chemist to assist with the translation, to ensure that there will be no automatic improvement of the chemistry, as could occur spontaneously if a chemist translates chemistry.
31. Mammino, L. *Nuova Secondaria* **1996**, *10*, 75–76.
32. Love, A.; Mammimo, L. *Zimb. J. Educ. Res.* **1997**, *9* (1), 1–17.
33. Mammino, L. *Enseñ. Cienc.* **2002**, *20* (1), 167–173.
34. Castro, E. A. *Panam. Newslett.* **1995**, *8* (4), 8–9.
35. Kovac, S. *J. Chem. Ed.* **1999**, *76* (10), 1399–1402.
36. Mammino, L. *ALDEQ* **2006**, *XXI*, 241–245.
37. I have been particularly lucky in this regard, learning how my mother tongue expresses logical relationships first through the study of its grammar (primary school) and then through the study of Latin, the *ancestor* of Italian (Middle School). This provided a sound basis, on which it became particularly easy to learn how other languages express logical relationships, as the learning becomes tantamount to fitting new tools (the ways of expressing the relationships) into sets of familiar *boxes* (the types of relationships). An example illustrating the *familiar boxes* and *new tools* concepts is given in Reference (*38*).

38. A wide-range illustrative example is offered by a type of relationships that sometimes students find difficult even within the mother tongue, if there has not been specific training while studying the mother tongue within formal instruction: the relationships involving hypotheses. These relationships have a fundamental role in science, above all in its interpretation components. They involve three major aspects: reality (*if A is true, B is also true*), possibility (*if A is true, B may also be true*) and unreality (*if A had been true, then B would also have been true* – what implies that neither A nor B are true). While considering the statements A and B in terms of *being true* is more closely linked to a philosophical discourse, a more accessible view for students is in terms of happenings, i.e., respectively: *if A occurs, B also occurs* (reality); *if A occurs, B may also occur* (possibility); and *if A had been true, then B would have occurred* (unreality). The statement *If a system is isolated, no heat exchange is possible between the system and the surroundings* offers an example of reality; the statement *If a system is open, it may exchange mass and energy with the surrounding* offers an example of possibility (the possibility of exchanging is there, but it is not necessarily happening all the time); the statement *If Thomson's model of the atom had been true, all the α particles in Rutherford's experiment would have behaved in the same way* offers an example of unreality (the α particles did not behave in the same way and, therefore, Thomson's model of the atom could not be true). It is not easy to understand these conceptual differences through a language that a learner does not master adequately. On the other hand, once they become clear through the study of the mother tongue, it becomes easy to identify or learn the corresponding modes of expression in other languages (*37*). From personal experience, I learnt first that, in my language, they are mainly expressed by: «*se* + verb in the indicative mode» for the condition, and a verb in the indicative mode in the principal clause, for reality cases; «*se* + verb in the indicative mode» for the condition, and a verb in the conditional mode (or the presence of the modal verb *potere*) in the principal clause, for possibility cases; «*se* + verb in a past conjunctive mode» for the condition, and a verb in a past conditional mode in the principal clause, for unreality cases. This finds easy correspondence with the main ways of expressing the same relationships in English («*if* + verb in the indicative mode» for the condition, and a verb in the indicative mode in the principal clause, for reality cases; «*if* + verb in the indicative mode» for the condition, and the presence of the modal verb *may* in the principal clause, for possibility cases; «*if* + verb in a past conjunctive mode» for the condition, and a verb in a past conditional mode in the principal clause, for unreality cases) and in Spanish (where *si* is the equivalent of *if*). On learning Russian, where the verb system is not based on a plurality of modes, but on a complex use of prefixes and other tools, one has to learn largely different options: *если* is the conjunction equivalent to *if*, *може́т* and other forms of the same verb play the role of the English verb *may*, but the unreality case is expressed by the past tense followed by the particle *бы*. If somebody learning Russian already knows about the three possibilities in the hypothesis-consequence or condition-consequence relationships, it

is very easy to identify the correspondence between those possibilities and the modes of expressing them in Russian; if one has to learn to identify the meaning of a past tense accompanied by the particle *бы* without knowing about the condition-consequence relationship and its unreality option, it may be considerably more difficult to catch the meaning; if one were to read a scientific text without fully knowing about that relationship and how it is expressed in Russian, the meaning of the scientific text might be missed or misinterpreted. The last phenomenon is frequently observed (with English) among students at UNIVEN. In correspondence to unreality statements, students' way of expressing things may even suggest the vision of two different and co-existing worlds: e.g., a world in which Thomson's model is true and all the α particles in Rutherford's experiment behave in the same way, and a world in which not all the α particles in Rutherford's experiment behave in the same way and the nuclear model of the atom is true (*40*). Of course, this co-existence is the impression a reader gets from their way of wording; too often, those students' language-mastering is not adequate (even on verbal communication within interactions) for a teacher to be able to determine whether the problem is restricted to incorrect expression, or actually corresponds to a vision of two co-existing options.

39. Rubanza, Y. I. In *Rehabilitating African Languages*; Prah, K. K., Ed.; CASAS: Cape Town, South Africa, 2002; pp 39–51.
40. Mammino, L. *J. Mol. Struct.* **2005**, *729*, 39–45.
41. Gilbert, J. K. In *Visualization: Theory and Practice in Science Education*; Gilbert, J. K., Reiner, M., Nakhlekh, M., Eds.; Springer: Dordrecht, The Netherlands, 2008; pp 3–24.
42. Harre, R. *The Principles of Scientific Thinking*; Chicago University Press: Chicago, 1970.
43. Mammino, L. *ALDEQ* **2010**, XXV, in press.
44. Mammino, L. *ALDEQ* **1999**, *XII*, 70–73.
45. Mammino, L. *Proceedings of the 10th SAARMSTE Conference*: Durban, South Africa, 2002; pp 237–243.
46. Mammino, L. *ALDEQ* **1998**, *X*, 54–59.
47. Mammino, L. *ALDEQ* **1998**, *XI*, 123–126.
48. Mammino, L. *ALDEQ* **2001**, *XIV*, 25–28.
49. Mammino, L. *ALDEQ* **2002**, *XV*, 125–129.
50. Mammino, L. *SAARMSTE J.* **2001**, *5*, 29–40.
51. Mammino, L *ALDEQ* **2003**, *XVI*, 5–10.
52. Mammino, L. *ALDEQ* **2006**, *XXI*, 33–37.
53. Mammino, L. *ALDEQ* **1997**, *IX* (2), 178–189.
54. Mammino, L.; Cardellini, L. *Baltic J. Sci. Educ.* **2005**, *1* (7), 51–62.
55. Mammino, L. *ALDEQ* **2004**, *XVIII*, 46–52.
56. Mammino, L. *ALDEQ* **1998**, *XI*, 275–278.
57. Euclidean geometry provides valuable keys for this, as shapes can be perceived as concrete, but the way in which they are treated within geometry requires a good deal of abstraction abilities. This was broadly recognized in former secondary school syllabi in Italy, assigning such a role to geometry in those types of schools (like the humanities-oriented secondary school,

liceo classico) that focused more strongly on the development of thinking abilities.

58. Mammino, L. *Nuova Secondaria* **1995**, *5*, 84–86.
59. Mammino, L. *Chimica Aperta*; G. D'Anna: Florence, Italy, 2003.
60. Seepe S. In *The Language of Science*; Seepe S., Dowling D., Eds.; Johannesburg, South Africa, 2000; pp 40–51.
61. It is also a personal experience. At pre-university level, I learnt all the subjects in my mother tongue, and learnt English as one of the subjects. At university level, instruction was in the mother tongue, but most of the chemistry textbooks were in English, and the combination of the two features proved quite reflection-stimulating. At Ph.D. level, I used another language (Russian) that I had to learn on the spot.

Chapter 3

Multi-National Approach to Teaching and Learning

Liberato Cardellini*

Università Politecnica delle Marche, Via Brecce Bianche, 60131 Ancona, Italy
***libero@univpm.it**

Problem solving is universal to all cultures and backgrounds. This chapter provides a literature review of cognitive processes inherent to problem solving that can be applied to chemistry study in diverse cultures. The information processing model is shown as an example of how students incorporate chemical concepts into their own conceptual framework. The feasibility of lecture as a tool to develop problem solving skills is discussed. Pictorial representations of chemical concepts can be used to increase long-term conceptual understanding. All theories discussed in this chapter can be incorporated into teaching curricula in many diverse cultures.

"Teaching and learning are not synonymous; we can teach, and teach well, without having the students learn."

George Bodner

Prologue

Early in the '90s, I decided to improve the quality of my teaching, resolving to devote additional time to preparing lectures and the related didactic material. I then made a questionnaire that asked my students to evaluate my teaching. Notwithstanding my efforts in teaching and the huge amount of time that I gave to it, the overall benefit that students received, according to their opinions, was disappointingly low. This result was particularly distressing because didactics was my field of research and problem solving was an important skill for my freshman students in an engineering program. I did not surrender, and since that time, three events have changed my life and my teaching practice.

First, I was able spend some time at the University of Glasgow in the Alex Johnstone Centre for Science Education, a recognized place of excellence in chemistry education. I attended the lecture on problem solving that Johnstone delivered for the students, and while walking and talking with Alex in the Scottish highlands, I realized I had to change the chemistry curriculum I was giving to engineers. As required by the university curriculum, I was covering too much material in the time available.

Second, something also was wrong with the way I taught problem solving. I taught the solutions to the problems one step at a time, explaining the various steps, as it is usually done in academia. I may have taught an elegant explanation of the solution, the way my instructors taught me and the way one can find in the textbooks, *but it did not help students understand how to approach problems in a fruitful way*. At the time, I read an article by Richard Felder (*1*) who was both an expert in cooperative learning strategies and enamoured with education. Much as a father does, he helped me in my first hesitating steps applying the cooperative strategies with advice, suggestions and encouragements.

The third event began with the personal notes that a student left to me after the final exam. In this note, a student listed their notes taken during my lectures where they transformed my words into summaries and enriched by figures and graphics. I understood the potential importance of the artifact: it was a sort of "window into the mind" (*2*) that allowed me to understand how my students incorporated my lessons into their mental framework. From that point forward, I advised my students to draw concept maps. My desire to learn more made the search for personal contact with Joseph Novak, the inventor of concept maps, a necessity.

These three events, all interactions with colleagues in other countries, led me to research new types of teaching methods that could be applied at my institution in Italy. My research led me to a greater understanding of problem solving and cognition, things which are universal to many countries. I have also presented the results of my curriculum changes to chemistry instructors in Italy, Egypt, Pakistan, and Turkey. This chapter highlights some of the effective methods that were devised in classrooms in the UK and US, implemented in my classrooms in Italy, and taught in different countries with very diverse cultural backgrounds. Thus this chapter discusses good teaching practices that transcend cultural boundaries.

Introduction

In 1921, the *Journal of Educational Psychology* published the responses of fourteen experts to the question: "What is intelligence?" 57% of the responses indicated the attribute "Higher cognitive processes (e.g., reasoning, problem solving)". In 1986 Robert J. Sternberg and Douglas K. Detterman asked the same question to 24 leading psychologists, and 50% used the same attribute to define intelligence. This was the first among 15 different possible choices (*3*). Thus, problem solving is very important in many subjects in tertiary education. But before we can teach problem solving we must ask "What is a problem?"

In the literature there are several definitions of the term 'problem solving'. "A person is confronted with a *problem* when he wants something and does not know immediately what series of actions he can perform to get it" (*4*). "Whenever there is a gap between where you are now and where you want to be, and you don't know how to find a way to cross the gap, you have a problem" (*5*). There are many other definitions, but the point is that there is an obstacle or barrier in the path from problem to solution. In this way we have a criterion to distinguish between *exercises* and *problems*. "Problem solving is overcoming obstacles or barriers or bridging this gap by using INFORMATION and REASONING" (*6*). Exercises are problems that we can solve using rote memorization or practiced techniques. A more workable definition of problem solving is "the ability of students to answer correctly questions of the type commonly found in formal school or college examinations" (*7*).

From these definitions, it follows that problem solving is "thinking that is directed toward the solving of a specific problem that involves both the formations of responses and the selection among possible responses" (*8*). According to Frazer, the problems can be open or closed, in which "closed" problems have only one correct solution. Open-ended problems do not have unique, unambiguously correct solutions (*6*).

A useful distinction is between well-defined and ill-defined problems. A well-defined problem provides four different sorts of information: (*9*)

1. the *initial state* of the problem;
2. the *goal state*;
3. legal *operators* (things you are allowed to do in solving the problem);
4. *operator restrictions* which constrain the application of operators.

It is worth noting that most significant real-world problems are ill-defined and there is an interest studying how individuals solve complex and ill-structured problems, the use of diverse strategies to solve problems, and the development of a model for solving ill-structured problems (*10–13*).

Problems can be thought as having three components: the data, the goal, and the method for reaching the desired outcome. If one or more of these three components is missing or incomplete, we have a problem. Classification into one of eight different types of problems comes out from the possible permutations of these three components. In one situation, each of the three components is known: this type of problem is commonly called an exercise (*14*, *15*). The classification of problems is an attempt to make a distinction between exercises and problems, and to find a difference according to certain rules between the types of problems. Although interesting, the classification of problems is not very useful in practice because "Status as a problem is not an innate characteristic of a question, it is a subtle interaction between the question and the individual trying to answer the question" (*16*).

Cognitive scientists distinguish between declarative knowledge and procedural knowledge. Declarative knowledge is explicit knowledge of some object, event, definition, or idea and of which we are consciously aware. "Knowing that" is declarative knowledge while procedural knowledge is "knowing how". Procedural knowledge involves knowing how to do something; it is knowledge of a sequence of operations that can be applied to a class of tasks. "In understanding procedural knowledge we start with problem solving because it seems that all cognitive activities are fundamentally problem solving in nature" (*17*). Scientific conceptual knowledge and procedural knowledge are essential requisites for solving chemical problems.

Conceptual understanding helps the problem solver develop a meaningful representation of the problems, and also narrow the search for solutions, by matching the schema or "conditions" that represent the givens in the problem with a set of "actions" in procedural memory that are most likely to produce the correct solution of problems. A serious problem in chemistry is developing this conceptual understanding. Methods that can be applied in classrooms in different cultural settings are discussed in the following sections.

The Traditional Lecture

The main didactic concern of many teachers lies in the completion of the syllabus, while less emphasis is given to how much information and conceptual understanding is retained by students. According to the level of students' involvement in the process of learning, we can identify different educational strategies. In the lecture, the most common method of postsecondary teaching, the lecturer gives information and the student is mainly occupied in the effort to follow the explanation and take notes. Studies about its efficacy cast some doubts on its effectiveness for transmitting knowledge (*18*). Of about 5000 words heard in a 50 minute lesson, students noted down about 500 of them, and on average transcribed about the 90% of the information written by the lecturer on the blackboard (*18*).

Johnstone's research considers the working memory capacity and its dependence or independence from the field (Field dependence/independence, FD/FI). FD/FI is a psychological measure that highlights the ability of a person to distinguish the essential information from the rest of the message; it is a bit like splitting the "signal" from the "noise" (*19*, *20*). Moreover, the research considered the total number of units of information transcribed from the student. The information unit is defined as the smaller block of knowledge that has meaning as a separate statement. The analysis of the total number of recorded words, the total number of noted down units of information, and their completeness shows that there is a correspondence with the working memory capacity (*21*): students with a low working memory capacity recorded the same number of words noted by the students with high working memory capacity, but the quality and completeness of what gets written is lower. The correlation between FD/FI and the notes taken is similar, but if the units of information and their completeness are considered, we see that FI students record more complete and meaningful information. Ultimately, the traditional lesson favors the better students (*18*).

We may sum up the negative aspects of the lecture as an instructional tool by saying that it is not about the process of teaching and learning, but often it reduces to a stenography exercise; it has to do with the transmission of what Whitehead calls 'inert ideas,' "ideas that are merely received into the mind without being utilized, or tested, or thrown into fresh combinations" (*22*, *23*). "[...], a major problem with the lecture is that students assume a passive, non-thinking, information receiving role" (*24*). Even academically strong students who find it difficult to pay attention and stay interested for an entire hour or more. Others have reported specific suggestions on how to hold the attention of students (*25*, *26*).

The most important issue in teaching is what the students learn while they are in class with us. The motivation and the learning are related to the positive environment that as experts teachers, we are able to create. From the many literature findings it's safe to say that the personal involvement in the active construction of meaningful knowledge plays an essential role (*27–29*).

"Importantly, there appears to be two ways to acquire knowledge and to solve problems. One way is through shear repetition and/or via emotionally charged contexts. Repetition and emotion can "burn" new input into long term memory. ... Fortunately, there is a second way to learn. That way is to link new ideas with prior ideas. This connectionist (or constructivist) way of learning has several advantages, not the least of which is that learning is not rote. Instead it connects to what one already knows, (and) thus becomes much more useful in reasoning and problem solving" (*30*). In the next subsection, we consider some useful aspects of this approach.

Cognitive Structure and Knowledge

According to cognitive psychology, knowledge is the storage and organization of information in memory. Concepts, schemata, and ideas are represented by the nodes within a network. The nodes are connected to each other by links called associations (*31*). It is the working memory that transforms the information collected from the senses into knowledge.

The working memory is where understanding, thinking, making sense of information, reasoning, and problem solving take place while the long-term memory is where information, concepts, understandings, emotions and experiences are stored (*21*, *32*). "It is a limited shared space in which there is a trade-off between what has to be held in conscious memory and the processing activities required to handle it, transform it, manipulate it, and get it ready for storage in long-term memory store. If there is too much to hold, there is not enough space for processing; if a lot of processing is required, we cannot store much" (*33*). The working memory can retain information for a limited period of time, and has a capacity of about four to seven items, or chunks (*34–36*), while the capacity of long term memory appears to be infinite. The time needed for fixation of information in the long term memory is about 5 to 10 seconds per chunk (*35*).

Another model associated with information processing was developed within the neo-Piagetian theory of Pascual-Leone, the Theory of Constructive Operators (TCO) in which the concept of mental capacity was introduced (*37*). The theory was proposed as an account of individual differences in performance on cognitive tasks. Mental attention, the maximum number of schemes or discrete 'chunks' of information that the central processor (M) can attend to, or integrate in a single act, evolves with age as a function of maturation, increasing its power (*38*, *39*). According to Pascual-Leone, the information-processing ability of an individual represents "a reserve of mental energy (M-energy) that can be allocated to raise the activation weight of task-relevant schemes. This reserve of M-energy (maximum M-power) may be measured by the maximum number of different schemes that M can weight in a single mental act of centration" (*40*).

The first step in human processing and learning is perception. Perception is a filtration process by which we choose to attend to certain parts of our sensory input and to ignore others. This process is controlled by our long-term memory, where we decide on importance, interest and attention, based upon previous experience and knowledge (*41*). During a lecture, the working space is totally occupied with sending messages from the senses to the student's pen. This may explain why student's notes are often inaccurate: because they cannot process the incoming information and, therefore, fail to see mistakes (*18*).

There are many diagrammatic ways of illustrating the research evidence, but most are very similar. The model of Johnstone is used in Figure 1 because it was developed specifically for the context of chemistry learning (*42*).

Johnstone described learning like this.

> Long-term memory is a vast store of information inter-linked in huge association networks. The working memory compares the input with material stored in long-term memory looking for connections which will make sense of the input. If a fit is found, the new information is connected into the existing network correctly, increasing and enriching the complexity of it. If, however, a fit is not found, the information may be rejected and forgotten, or it may be "bent" to make it fit. This is where wrong ideas and misconceptions have their origin. It may find no point of attachment, but the student "parks" it in a space where it is difficult to retrieve it. It is not in his "filing system" and so gets lost easily. This is rote learning which students use before examinations to hold undigested information for a short period of time before it is lost because the working memory cannot find the path for finding the information again. This model plays a role in the way teaching and learning takes place, and the Ten Educational Commandments stem directly or indirectly from the information processing model (*43*, *44*).

Because of the limited room of the working memory, there will be a point where the effort of holding the information in memory conflicts with the information processing. The working memory is overloaded: as reported in Figure 2, many students fail to solve the problem. This means that there is a limit on the quantity of information that students can process which emphasizes the importance for teachers of evaluating the amount of information processing requested for understanding the lectures or to complete problems successfully.

We have to rely on experience to measure the cognitive load of the task but we can get an help from cognitive load theory (*45*). "The fundamental claim of this theory is that without knowledge about the human cognitive architecture the effectiveness of instructional design is likely to be random" (*46*). This is not meant to make the work of students lighter; our goal is to make the learning more efficient. The teaching of strategies and chunking devices helps students be more successful (*47*). The M-demand of a task can be reduced by training the students or by manipulation of the task's content (*48*). The current model of working memory is more complex compared with the Johnstone's model (*21*, *49*), but it is sufficient for our educational purposes and thus is a useful construct in chemical education.

The limitation of the working memory constitutes a severe obstacle on our capacity to solve problems. Reid states that "Science education research has battled for many decades in seeking solutions to the problems of learner difficulty which seem inherent in much of physics and chemistry as well as some areas of biology... [There is] abundant evidence that one key factor causing the learner difficulties lies in the limitations of working memory capacity" (*50*). Notwithstanding these and other constraints, it is possible to improve meaningfully the abilities of our students in problem solving.

Practice makes perfect: students can learn while engaged in solving problems (*51*), and the chunking process takes place with the repeated practice (*52*). The practice increase the amount of planning a person do while solving problems. The learning that result from this process increases problem solving efficiency, because of better planning, and frequently produced optimal solution (*53*).

Representation and Space of a Problem

Two other important and useful concepts in problem solving are the problem space and the problem representation. A *problem state* is a description of the elements of the problem: "the set of all the expressions that exist in the world of the problem at a particular time" (*54*). According to Anderson, "The actual reference of *state* is ambiguous. It could mean either some external state of affairs or some internal coding of that state of affairs. Newell and Simon, with their emphasis on problem solving by computer, typically took it to mean the internal coding" (*55*). A connected construct is that of a problem solving *operator*. An operator transforms one state into another. These two concept together define the *problem space*.

"The problem space refers to the problem solver's internal representation of: *Initial state* – in which the given or starting conditions are represented, *Goal state* – in which the final or goal situation is represented, *Intermediate problem state* – consisting of states that are generated by applying an operator to a state, and *Operators* – the moves that are made from one state to the next" (*56*). "The **problem space** is all the possible solutions to the problem that have occurred to the problem solver. The problem may have other solutions. However, if the problem solver is not aware of these solutions, they are not included in the problem space" (*57*). The elements that constitute the representation of a problem are also included in the problem space. "The problem space consists of states of knowledge and operators that can be applied to elements in the space to produce a new state of knowledge" (*58*). Connected with these ideas there are weak problem solving methods, that can be be applied to a wide range of domains. "They are called weak because they do not take advantage of domain characteristics" (*59*). One of these weak methods called means-ends analysis, can easily and usefully be applied in the solution of chemical problems (*60*).

So we are arrived at the representation of the problem. "A *problem representation* is a cognitive structure corresponding to a problem, constructed by a solver on the basis of his domain-related knowledge and its organization" (*61*). The representation consists in the solver's interpretation and understanding of the problem. According to the information processing model, the representation of the problem generally depends on the information contained in the long-term memory. Because of this, different students use different representations of the problem. The manner in which a problem is initially described is crucially important since it can determine how, and often whether, it will be solved in a correct way.

In the mentioned study (*61*), a meaningful difference between experts and novices in solving physics problems has been found. While the experts spend time in qualitative analysis of the problem, novices start with writing equations. Experts also tend to categorize the problem according to the laws of physics, while students categorize the problem according to some superficial entities and descriptions contained in the text of the problem (*61*). While the expert generates a physical representation of the problem, the novice often uses a process of direct syntactic translation (*36*).

Research has shown that students in chemistry found difficulties with the different levels of representation of the particulate nature of matter (*62*, *63*). The information obtained from these studies has serious implications for the teaching of chemistry, because the lack of understanding of the particulate nature of matter makes it difficult, if not impossible, to solve related problems correctly. In spite of the well-known difficulty that students have moving among the microscopic, symbolic, and macroscopic level of description of the chemical world, "no more than 1% of the questions in the analyzed textbooks target this skill" (*64*).

One way to reduce the overload of the working memory and to make the reasoning visible, is to make a pictorial representation of a problem. "Chemistry students find solution problems involving dilutions and additions of solute very difficult. How can these problems be solved successfully without picturing how the addition of the solute to the solvent causes the solute particles to became closer together and the solution more concentrated?" (*65*). It is important to consider and understand both the precise meaning of the terms and the specific physical situation described in the problem before solving it (*66*). Sometimes is difficult to distinguish a logic error from a chemistry error (*67*).

The use of pictorial problem-solving networks makes it easier for students to determine, for example, the empirical formulas of oxides and reduce their perceived difficulty of the subject (*68*). A similar approach was used in titration calculation problems by the same author (*69*): because of the additional complexity of the problem, the instructor suggested identification of the solution involved in the titration. This decreased the mental energy required to solve

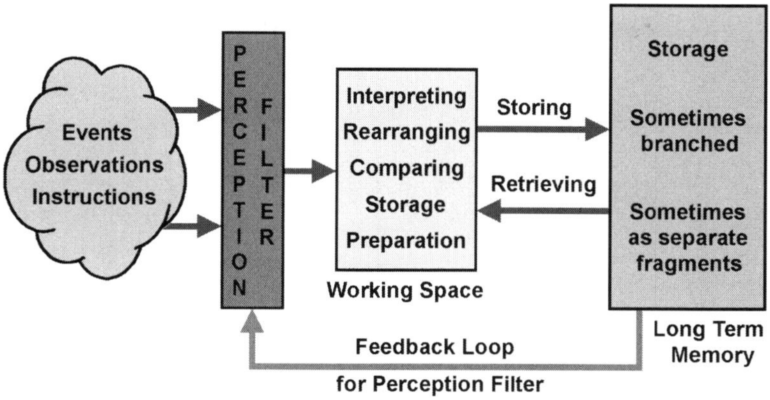

Figure 1. Information Processing Model (after Johnstone, 1997). Used with permission from the Journal of Chemical Education, Vol. 74, No. 3, 1997, pp. 263-264. Copyright ©1997, Division of Chemical Education, Inc.

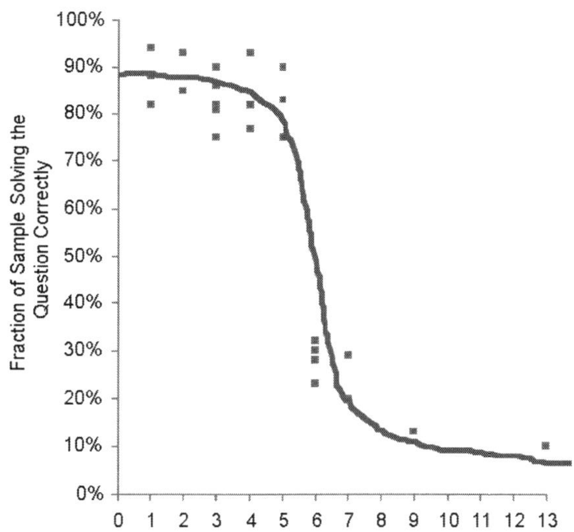

Figure 2. Complexity of question vs. fraction of students that solve the question correctly. Used with permission from the Journal of Chemical Education, Vol. 74, No. 3, 1997, pp. 263-264. Copyright ©1997, Division of Chemical Education, Inc.

the problem because of the reduced information content of the problem, and improved the confidence of the students in their ability to solve the problem (*69*). The pictorial problem-solving networks are used as a cognitive support and aid in the cognitive development of the students.

Some students are unable to understand the difference between atom and molecule and the stoichiometric relationships involved in a chemical equation (*70*). A useful tool is to use pictorial representations of the problem in this way:

H_2O

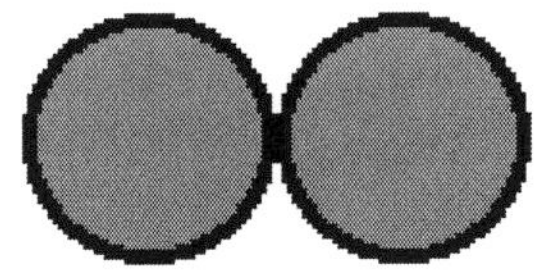

H_2

In this way, the stoichiometric relationship is visible: 1 mol $H_2O \equiv$ 1 mol H_2. If the students are asked to comment on, and verify, any mathematical calculations, then to solve a second example in which the question is turned around according to the strategy Turn it Around (*71*), and again to comment and verify any mathematical calculations, chances are that our students will learn correctly the logical relationships.

Individual Differences in Solving Problems

According to the Chambers Dictionary, a problem is "A matter difficult of settlement or solution: a question or puzzle propounded for solution ... a proposition in which something is required to be constructed, not merely proved as in a theorem: a source of perplexity" The word 'problem' has negative connotations and evokes something doubtful that must be solved; a choice among some options. Still, problem solving plays a valuable role in education, and Watts enumerates eight reasons that justify his use (*72*).

Chemical problem solving constitutes a stumbling block for many students in introductory courses. Silberman reported the results of a Delphi study with the aim of discovering the reasons for problem-solving difficulties and suggestions for improving these abilities. Interestingly enough, the author's conclusion was that in order to improve student problem solving performance it is necessary to increase students' motivation and to use material that will expand their abstract reasoning ability (*73*).

In order to solve chemical problems successfully, a strong background knowledge of chemistry is essential, as well as the knowledge of chemical strategies in problem solving (*74*). However, it has been shown that the possession of chemical knowledge and the knowledge of strategies and skills are not sufficient to solve a problem if confidence arising from previous experiences of successful problem solving is missing (*6*). But also a clear picture of the problem in necessary. "Unsuccessful problem solvers may be unable to perceive a plan to solve the problem because they lack confidence and become confused when they read an unfamiliar term or unfamiliar unit, or when they are confronted with an unusually long problem (*75*).

Kempa and Nicholls (*7*) used chemistry achievement and word association tests to explore the relationship between students' cognitive structure and their problem-solving abilities in the context of chemistry. They found that the cognitive structures of good problem-solvers are more complex and contain more associations than those of poor problem solvers. The strength of links among different concepts seems important in determining problem solving behavior. It was also revealed that the deficiencies in the cognitive structures of poor problem solvers appear predominantly for abstract concepts. Although this study used problems mainly of an exercise or algorithmic nature, its findings may also apply in more open-ended problems.

The first step in successful problem solving is to understand the meaning of the text of the problem. Cassels and Johnstone studied the effect of the language on student performance and found that when the keyword or the terms in questions were simplified, a greater number of students were able to solve the problem correctly. The authors attributed this finding to the idea that reading is more than a visual process; it involves the interaction between the information received with the visual system and the information already available in the long term memory (*76*). The knowledge of scientific terms is a necessary, but not sufficient requisite for success in solving problems. It was found that even if students have a basic understanding of chemical terms, but do not recognize the relationships between them, they are unable to apply their knowledge. (*77*). Many 14 year-old students have difficulty using logical connections, and language difficulties can also be a handicap in the assessment, because the demonstration of reasoning abilities takes place through the use of linguistic material (*78*). These studies demonstrate that language influences the thinking process activated to answer questions or solve problems. An effective strategy for reading comprehension is to read the problem again by *first* reading the question and then reading the rest of the text of with the question in mind (*79*).

Individual differences in problem solving are important because "any proposal for a "general" theory of rule induction—or of problem solving, for that matter—must account for the fact that there are wide individual differences in behavior in these task environments, and that these differences can be enlarged further by relatively small changes in the conditions under which the tasks are performed" (*80*). Individual differences in solving problems were found between persons of different abilities in selecting relevant from irrelevant information (*81–83*). The ability to separate 'noise' from 'signal' can be measured and interpreted using the field dependence/independence construct developed by Witkin (*19*, *20*, *84*). Research has tried to correlate some cognitive variables, such as formal operational reasoning, working memory capacity, Witkin's cognitive style, prior knowledge, concept relatedness and idea association, to science achievement and problem solving ability (*47*, *49*, *81–83*, *85–100*). Suggestions from research are that instructional methods should take into account the general strategies and methods of problem solving, thus providing a tool to increase reasoning skills in the problem solver.

There are other differences in performance between experts and novices in problem solving that deserve to be considered. Experts are able to solve complex problems much faster and more accurately than students. In solving physics problems the novice expresses a lack of confidence and, because of that, sometimes abandons a solution attempt even if it they are correct. In the development of the solution, the expert "moves from the problem statement to a representation of the physical situation, and from that representation to a set of equations" (*101*). With the greater experience the expert has developed a sort of 'physical intuition' that makes everything easier. Because of the limitation of the working memory, the developed ability of the expert in chunking the information and in using compiled cognitive procedures makes a difference. "The expert apparently had stored directly (perhaps as a production) an entire procedure for obtaining a desired value from related known values" (*102*). It was found that novices solved most of the problems by working backward, while the expert usually worked forward from the givens to the unknown (*36*). This different direction of the inference has been questioned (*103*). While the expert generates a physical representation of the problem, the student often uses a process of direct syntactic translation (*59*, *104*, *105*).

Cooperative Learning and Concept Maps

I wish to know that my students are engaged in learning during the course and an useful tool for this purpose is concept mapping. This tool help students lo learn more meaningfully and to help them to reflect on what they are learning. In my own classes, I have had success putting my students to work on problems in teams according to cooperative learning conditions. Cooperative learning is an instructional approach to group work that involves students working in teams toward a common goal. An instructor that wishes to use this method must be prepared and know how to use it. The reward comes from a greater active participation of students in their learning.

A learning exercise can be classified as cooperative learning if five elements are present (*106*, *107*):

1. *Positive interdependence.* Team members must rely on one another to accomplish common goals. Students take responsibility for their own learning and for the learning of their teammates.
2. *Individual accountability.* Members are held accountable for (a) doing their share of the work and (b) mastering all material.
3. *Face-to-face interaction.* Although some of the group work may be parcelled out and done individually, some must be done by members working together.
4. *Appropriate use of interpersonal skills.* Team members are encouraged, practice, and receive instruction in leadership, decision-making, communication and conflict management.
5. *Self-assessment of group functioning.* Teams periodically reflect on what they are doing well as a team, what they could improve, and what they will do differently in the future.

The best answer to the question, "What is the most effective method of teaching?" is that it depends on the goal, the student, the content, and the teacher. But the next best answer is, "Students teaching other students" (*108*). The method has the advantage of encouraging students to interact with their peers, and cognitively providing opportunities to explain to peers and oneself the concepts and the steps in problem solving. "Students find a great deal of meaning in science courses when their knowledge is constructed during productive, small, cooperative group activities" (*109*). This may be because students retain almost all of what they say, or what they say as they do something (*110*).

A concept map (Figure 3) is a graph consisting of nodes representing concepts and labelled lines denoting relationships among the concepts (*111*).

Because the making of summaries and maps are time-consuming activities, if students work on them, we have some evidence that our students are studying and reflecting on the material, especially if from time to time we call someone to the blackboard asking to answer questions with the use of her or his artifacts. Concept maps are used virtually in all learning settings, in education as well as in business (*112*). Concept maps are common in chemical education too, and interested readers can find articles in journals such as: Journal of Chemical Education, European Journal of Science Education, International Journal of Science Education, Journal of Research in Science Teaching, School Science Review, and Science Education.

Concept maps and summaries are tools that fit well in an active learning environment. From the students' point of view, there is a continuum between these two cognitive tools and some students' artifacts are difficult to assign to one of the two categories. This is not a problem because the tools are instrumental to a better learning. Students are encouraged to explain the concepts in the summaries and to argue in a way that is useful in defending their knowledge.

In the first lessons, often in the very first, I instruct the students on how to draw a concept map, according to the four criteria suggested by Novak and Gowin (*113*). Each student receives the handouts with some examples drawn by students of previous years. Students are encouraged to draw the maps and to make meaningful summaries for three related purposes: to make explicit the structure of their knowledge; to check their knowledge as it grows; and to have a meaningful summary to go over in preparation for an written and oral examination.

After every new lesson, students are requested to hand over the concept maps. I look at many of them, and in the following lesson, the maps are returned to the owners, sometimes with advice on how to improve them. When appropriate, some suggestions and question marks (using a pencil) are reported, such as when a student uses incorrect words to link concepts or the wrong concepts are listed. Students can use the maps during the written and oral exams. To avoid the use of these artifacts to copy out formulas or solutions of problems, the oral examination begins with the qualitative explanation of the written exam. After the exam, almost all the students leave the maps with the teacher, and over the years I collected about 15,000 of them.

Chemistry: An Alien Language?

For many reasons, chemistry is recognized to be a difficult topic and many reasons, explanations, or aspects of chemistry that make the topic difficult have been discovered by research and reported in the literature (*114–116*).

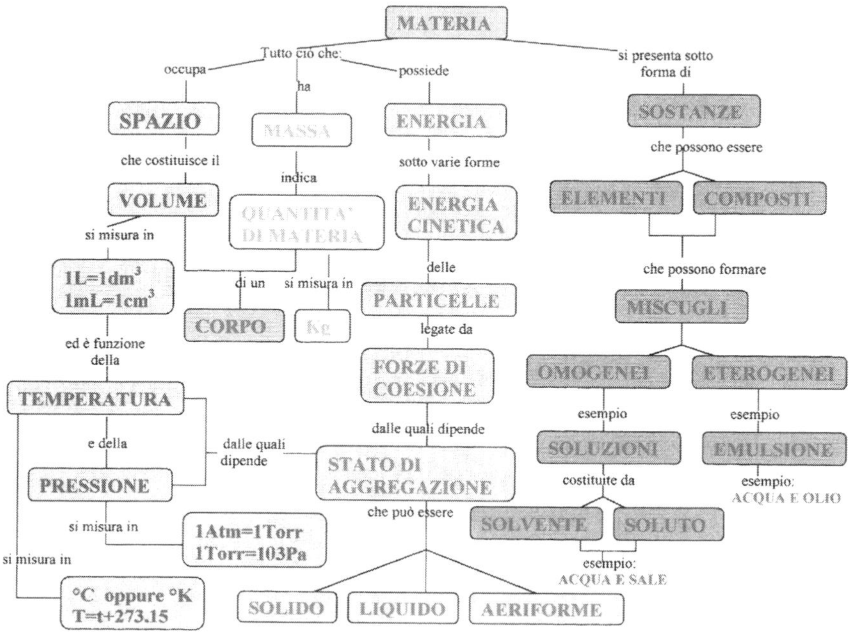

Figure 3. Example of a concept map developed for the concept 'matter'. The use of different colors for related concepts and relationships helps to systematize the content.

Unfortunately, many instructors who teach chemistry are not completely aware of these difficulties, and because of that, discourage students from enjoying chemistry. The majority of my students arrive at the university with a poor interest for the subject and even poorer knowledge of chemistry. The dearth of chemical knowledge is illustrated by student answers to a problem I call Students-and-Professors.

A questionnaire was distributed in two of my previous courses to about 400 freshman engineering students, where students had to select the correct equation that defined the following statement (*104*).

> There are six times as many students as professors at this university." (In the equations, S stands for the number of students and P stands for the number of professors.)

The logic at the bottom of this problem is the same as in many stoichiometric problems (*67, 117*). Two hundred and sixty students responded to the questionnaire and about 30% of them choose the right equation. About the same fraction (N = 84 responses) solved correctly the simple stoichiometric problem:

> "10.00 g of Na_2CO_3 reacts with 10.00 g of HCl. One of the reagents is completely consumed. Calculate the grams obtained of every product, explain your reasoning, and outline a method to verify your results."

Some remember a rote-learned algorithm and can solve the first part of the problem, but no one could explain their logic and quite few verify their results. Other assessments, such as The Programme for International Student Assessment (PISA) show that this data, with some caution, can be generalized to the whole country.

If we try to understand this state of affairs, part of the culprit is the educational system that has no National standards, leaving the teacher to decide on the level of student preparation. Because teachers have a vested interest in showing good results, in many cases students can arrive at the university without having never had a significant experience of studying.

The same thing happens at the university: it is the teacher who decides who passes the exam and who does not. This is true in Italy and in much of the Europe. In some countries, including Ireland and the UK, there is often second marking by a colleague. In the past, the university exam consists of a written and an oral examination. Because of the large numbers of students involved in university courses, many colleagues use written exams (usually quizzes), as they are a more objective method of evaluation and are more easily assessed. This also happens in other countries, including Poland and Portugal. My general chemistry course, taught by the engineering faculty, still consists of two written exams (a mid-term and a final exam) on stoichiometric problems, along with an oral examination. Students who fail to pass the written examination can repeat it when they are prepared for the oral examination. Six exam sessions during the academic years are quite common in Italy.

The absence of a common objective standard has advantages and disadvantages: the disadvantages can be the students' preparation. Because we want to be considered good teachers, we may use a low standard, so that weak students can easily pass the exam. The advantage is that the teacher does not teach with the goal of preparing their students to pass the test. This means that some motivated and highly professional teachers do a very good job and their students are prepared to work seriously for their learning. These students arrive at the university with the desire to learn meaningfully. It is the job of the university teacher to encourage such students. On the contrary, if we are able to interest them and give them the chance to show their value, they will become an example for the whole class and improve the learning environment of our course. Even many of the weaker students are ready to be engaged, as long as we, as teachers, are able to interest them in our subject.

A Viable Approach

Many high school students feel anxious when they arrive at the university because they discover that they do not know how to study seriously and find it difficult, if not impossible, to start studying methodically at 20 years of age. Asked about their expectations and purposes in the very first days of the courses, many confess that they never find chemistry interesting, or they have forgotten what they have studied. As an excuse, several students adduce that the high school's instructor was not interested or properly trained in chemistry. One problem in

Italy is that it is not necessary to have a degree in chemistry to teach chemistry at secondary school level. A more serious problem is that many chemistry instructors lack the enthusiasm for, and interest in, teaching chemistry.

Because of this background, and in order to obtain good results in education, it is crucial to create a supportive learning environment, especially during the first days of school. "In supportive environments teachers expressed enthusiasm for learning, where respectful, used humour, and voiced expectations that all students would learn" (*118*). Because motivation is more a process than a product, I try to attract and interest my students in learning those things that I deem worthwhile, by using logical problems. Some of the problems are easy, so each student can and will solve them. I never miss an occasion to establish a cordial relationship and to praise students by email or in class who excel in something related to learning or problem solving. I encourage the students to solve more problems and I give positive feedback that all of them will do well. In this friendly and supportive environment, several students, especially the intrinsically motivated students (*119*), start to solve problems, make summaries and draw concept maps. These students are the secret weapons for winning the engagement of the large part of the class.

An environment that stimulates interest and motivates students to acquire knowledge can minimize what has been called the problem of inert knowledge (*22*): knowledge, although seemingly available, is often not used for problem solving. This type of learning environment fits the principle of similarity between the learning and the application contexts (*120*): an environment in which many chemical problems had to be solved and the corresponding knowledge had to be acquired and applied.

I start the course greeting the students and shaking hands with many students in large classes. I try to learn their names, and when I call someone to the blackboard, or ask a question, as often I do during lectures, always I ask the name, if I do not know it. I want to signal to my students that they are important to me. I always try to establish a supportive relationship, which includes going to the cafeteria with some students, or replying thoughtfully to their e-mails.

Because problem solving abilities are so important in an engineering faculty, about half of my class time is spent on group problem solving. Students in each team have to solve a problem using the cooperative learning method. Students are asked to write their name on the solution sheet and state the role that they assumed in the cooperative learning group, with the roles rotating in each new exercise. The students know that one of them will be randomly called to the blackboard to present and explain the solution. The students know that the problems in the written exams are chosen between the many problems they will solve in teams or as homework. This provides a high level of motivation to solve the problems. They have to agree on one solution, and every team member must be able to explain the strategy used to solve the problem and to verify their solution.

I have developed some general strategies which I call mind tools; the roll with a filling metaphor, the Hansel and Gretel's method, and the disco analogy that I use to explain the solution of stoichiometric and equilibrium problems. In the roll with a filling metaphor, the logic of making a sandwich is used to suggest a strategy for solving stoichiometric problems. In the Hansel and Gretel's method, the story of

Hansel and Gretel is used to present idea of working backward from a solution to the beginning. In the disco analogies, familiar situations in a disco are used to teach how to solve equilibrium problems (*121*). A logic procedure is used to explain ionic equilibrium calculation (*122*). I ask my students to solve cooperatively even problems on new topic. While they solve the problems in groups, I wander around and look over the shoulders of some teams, making comments or suggestions, and also control the time spent on the task. I give clues for helping them reason and continue to cooperate, but I do not explain how to solve the problem.

Many students wish to have personal explanations, advice, and clues to solving a problem they find difficult, so I receive hundreds of e-mails per course and I reply to each. Before starting each new lecture, I collect the students' homework: problem solutions, concept maps, and summaries. I subsequently correct each solution, noting the solution times and whether the students explained their steps, used proper units and had the correct number of significant digits, and I then verify the results. I take note of each student's work. This approach is considered radical because the teacher's engagement is dictated by the students' cognitive difficulties.

In my last course 39 out of 50 students passed the final exam. Each of them solved, on average, 142 individual problems (ranging from 33 to 433) and I collected 715 maps and summaries from 33 of them. To draw concept maps and to make meaningful summaries is a time consuming activity, so some students were not interested in these cognitive tools. In this course I corrected about 6,500 solutions of problems: few students dropped and some have still to pass the exam. In Italy a student has the right to try the exam six times in an academic year. If we wish to improve the learning standard, we have to be involved in the students' learning process. If we have the resources to correct the homework problems and give feedbacks to our students it is better, because it increases the students' motivations.

An important issue is how to deal with the errors made by students while solving problems. The student needs to understand what went wrong with the line of reasoning. According to Ernst von Glasersfeld, "When students make what the teacher considers to be an error, the teacher should try to find out what train of thought led the student to make that statement" (*123*). When I found an error in a students' homework, I inform the student and ask him or her to solve the problem again. Dealing in this way with the errors is productive: as the course proceeds, I find fewer and fewer errors in the homework problem solutions, and when I examine the solutions I find evidence of students correcting themselves.

Epilogue

Does this approach to teaching and learning interest colleagues of other cultures? Well, it can. On November 2008 I was involved to the 10 day International Workshop on Systemic Approaches to Teaching and Learning Chemistry (SATL: http://www.satlcentral.com, accessed April 2010), at the University of Karachi in Pakistan; about 40 colleagues attended the workshop. SATL is an approach grounded in constructivist theory, developed by Farouk

Fahmy, from Ain Shams University in Cairo, Egypt, and Joseph J. Lagowski, The University of Texas at Austin, Austin, Texas, USA, with the aim of making the learning of chemistry and other subjects more meaningful (*124*). This approach is quite popular in Egypt, where an great number of pre-college science educators have received formal training in SATL sponsored by the Egyptian Government (*125*).

Deep learning, and problem solving in particular, requires the interest, motivation and engagement of students. In Shulman's *Table of Learning* (*126*), the first step is Engagement and Motivation. Learning begins with student's interest and engagement in the subject, which in turn leads to knowledge and understanding. So the real question is: how can we teachers interest and engage our students in chemistry? My students become interested in chemistry because they feel their learning is important to me. They know my engagement and my work for them so even if I do not say it explicitly, they know I have high expectations for their learning: our example establishes a certain learning environment. If we are seriously engaged in the teaching of chemistry and with their learning, many students feel somehow obliged to comply to the same standard. This means we take learning seriously because, as Shulman noted (*127*), we take something seriously when we profess it. And we profess something when we make it public. And we take learning seriously if we take the learning of our students seriously. Because of the recurrence, questions, and advice requested, my colleagues know my interest for the students' learning: the germ of a community of teachers is good for our students' engagement (*128*).

The knowledge of the general strategies in problem solving, the difficulties students find in solving stoichiometric problems, some knowledge about our mind work, is a great advantage for helping our students to become better problem solvers and to get them interested in chemistry. If I have some success in improving my students' abilities in problem solving, it is because of the knowledge of the colleagues' studies on the subject. At times, when I have to reply to students' questions, I regret not knowing more. There are some general problem solving strategies that can be usefully applied to solve stoichiometric problems. If our students learn how to solve problems, they will develop real self-esteem. Indeed, it may be that confidence is developed from experience, especially successful experience. A person may be willing to take cognitive risks, given a background of successful experience.

Years ago, I looked for a correlation between the chemistry grade and the academic degree grade of students of our engineering college. My rationale was that a student with a greater self-esteem in his cognitive abilities will have success in other subject also, so he can get a better grade at graduation. The grades in Italian universities go from 18 to 30/30 cum laude. I considered the gamut, from less than or equal to 21/30, and greater than or equal to 27/30. Unfortunately, students with a modest grade in chemistry had a great chance to get a modest graduation grade. But the brilliant chemistry students received a very good grade at graduation: 93.6% of my students that get a grade equal or greater to 27/30 in chemistry, graduate with 110/110 or 110 cum laude. Data of my colleagues' students showed the correlation of 60-70%. The study is based on a statistics of about 5,500 graduated students.

Even if the large majority of my students arrive at the university with no interest in chemistry or, detesting it, they leave the course with a different perception of the subject. It seems that this approach is appreciated by many students, according to their evaluation of the course, reported in the official university site at the URL: www.univpm.it/docenti/Cardellini (allegati; accessed April 2010). Even if chemistry is a minor topic in their professional formation, I often receive messages from them after the exam or from old students, now successful professionals. Many students thank me because they enjoyed solving problems and find pleasure in this activity. I had few students that ask me more logic problems after passing the exam. It is not certainly whether creativity can be taught, but whether it can be fostered by the right environment. Course after course I have more and more students who solve difficult problems in an original and creative way (*129*).

In my search for making my teaching more interesting, I think I have found two guiding principles: one is from Richard Zare: "Inspiration is more important than information." Another one comes from Dudley Herron: "The real question is how one can get students interested in learning—more correctly, interested in learning those things that adults deem worthwhile. Seduction, I think" (*130*).

Acknowledgments

I am grateful to the editors, Charity Flener Lovitt and Paul Kelter, who offered suggestions for improving an early draft of this paper.

References

1. Felder, R. M. *J. Chem. Educ.* **1996**, *73*, 832–836.
2. Shavelson, R. J.; Ruiz-Primo, M. A.; Wiley, E. *High. Educ.* **2005**, *49*, 413–430.
3. Mayer, R. E. *Thinking, Problem Solving, Cognition*, 2nd ed.; W. H. Freeman: New York, 1992; p 327.
4. Newell, A.; Simon, H. A. *Human Problem Solving*; Prentice Hall: Englewood Cliffs, NJ, 1972; p 72.
5. Hayes, J. R. *The Complete Problem Solver*, 2nd ed.; Erlbaum: Hillsdale, NJ, 1989; p xii.
6. Frazer, M. J. *Chem. Soc. Rev.* **1982**, *11*, 171–190.
7. Kempa, R. F.; Nicholls, C. E. *Eur. J. Sci. Educ.* **1983**, *5*, 171–184.
8. Solso R. L. *Cognitive Psychology*, 4th ed.; Allin and Bacon: Needham Heights, MA, 1995; p 440.
9. Kaney, H. *Problem Solving: A Cognitive Approach*; Open University Press: Milton Keynes, U.K., 1986; p 20.
10. Simon, H. A. *Artif. Intell.* **1973**, *4*, 181–201.
11. Jonassen, D. H. *ETR&D* **1997**, *45*, 65–94.
12. Fernandes, R.; Simon, H. A. *Policy Sci.* **1999**, *32*, 225–245.
13. Voss, J. F. *Argumentation* **2005**, *19*, 321–329.

14. Johnstone, A. H. In *Creative Problem Solving in Chemistry*; Wood, C., Sleet, R. Eds.; The Royal Society of Chemistry: London, 1993; pp iv–vi.
15. Johnstone, A. H. *U. Chem. Ed.* **2001**, *5*, 69–73.
16. Bodner, G. M. *J. Chem. Educ.* **1987**, *64*, 513–514, 513.
17. Anderson, J. R. *Cognitive Psychology and Its Implications*, 4th ed.; Freeman: New York, 1995; p 237.
18. Johnstone, A. H.; Su, W. Y. *Educ. Chem.* **1994**, *31*, 75–79.
19. Witkin, H. A. *Psychological Differentiation: Studies of Development*; Wiley: New York, 1974.
20. Witkin, H. A.; Goodenough, D. R. *Cognitive Styles, Essence and Origins: Field Dependence and Field Independence*; International Universities Press, Inc.: New York, 1981.
21. Baddeley, A. *Science* **1992**, *255*, 556–559.
22. Whitehead, A. N. *The Aims of Education*; Macmillan: New York, 1929. http://www.anthonyflood.com/whiteheadeducation.htm (accessed April 2010).
23. Renkl, A.; Mandl, H.; Gruber, H. *Educ. Psychol.* **1996**, *31*, 115–121.
24. McKeachie, W. J. *Teaching Tips: Strategies, Research, and Theory for College and University Teachers*, 9th ed.; D. C. Heath and Co.: Lexington, MA, 1994; p 68.
25. Chiappetta, E. L.; Koballa, T. R. *Science Instruction in the Middle and Secondary Schools: Developing Fundamental Knowledge and Skills for Teaching*, 6th ed.; Pearson: Upper Saddle River, NJ, 2006; p 117.
26. Bodner, G. M. *J. Chem. Educ.* **1986**, *63*, 873–878.
27. Bransford, J. D.; Brown, A. L.; Cocking, R. R. *How People Learn: Brain, Mind, Experience, and School*, Expanded ed.; National Academy Press: Washington, D.C, 2000.
28. Bain, K. *What the Best College Teachers Do*; Harvard University Press: Cambridge, MA, 2004.
29. Novak, J. D. *Learning, Creating, and Using Knowledge: Concept Maps as Facilitative Tools in Schools and Corporations*; Routledge: New York, 2010.
30. Cardellini, L. *Am. Biol. Teach.* **2005**, *67*, 140–148.
31. Reisberg, D. *Cognition: Exploring the Science of the Mind*; Norton: New York, 1997; p 258.
32. Reid, N. *Res. Sci. Technol. Educ.* **2009**, *27*, 131–137.
33. Johnstone, A. H. *J. Chem. Educ.* **1997**, *74*, 262–268, 263.
34. Miller, G. A. *Psychol. Rev.* **1956**, *63*, 81–97.
35. Simon, H. A. *Science* **1974**, *183*, 482–488.
36. Larkin, J.; McDermott, J.; Simon, D. P.; Simon, H. A. *Science* **1980**, *208*, 1335–1342.
37. Pascual-Leone, J.; Smith, J. *J. Exper. Child Psychol.* **1969**, *8*, 328–355.
38. Pascual-Leone, J. *Acta Psychol.* **1970**, *32*, 301–345.
39. Pascual-Leone, J. In *The Future of the Cognitive Revolution*; Johnson, D. M.; Erneling, C. E., Eds.; Oxford University Press: New York, 1997; pp 75–101.
40. Pascual-Leone, J.; Goodman, D.; Ammon, P.; Subelman, I. In *Knowledge and Development*; Gallagher, J. M., Ed.; Plenum: New York, 1978; Vol. 2, pp 243–289, p 266.

41. Ausubel, D. P.; Novak, J. D.; Hanesian, H. *Educational Psychology: A Cognitive View*, 2nd ed.; Holt Rinehart and Winston: New York, 1978.
42. Johnstone, A. H. *J. Chem. Educ.* **1997**, *74*, 262–268.
43. Johnstone, A. H. Thinking about Learning. http://wwwcsi.unian.it/educa/pedagogia/ahj-10co.html (accessed April 2010).
44. Johnstone, A. H. *J. Chem. Educ.* **1997**, *74*, 262–268, 263–264.
45. Sweller, J. *Cog. Sci.* **1988**, *12*, 257–285.
46. Schnotz, W.; Kürschner, C. *Educ. Psychol. Rev.* **2007**, *19*, 469–508, 470.
47. Johnstone, A. H.; El-Banna, H. *Educ. Chem.* **1986**, *23*, 80–84.
48. Niaz, M. *J. Chem. Educ.* **1989**, *66*, 422–424.
49. Niaz, M.; Logie, R. H. *Oxford Rev. Educ.* **1993**, *19*, 511–525.
50. Reid, N. *Res. Sci. Technol. Educ.* **2009**, *27*, 245–250, 245.
51. Anzai, Y.; Simon, H. A. *Psychol. Rev.* **1979**, *86*, 124–140.
52. Newell, A.; Rosenbloom, P. S. *Cognitive Skills and Their Acquisition*; Anderson, J. R., Ed.; Erlbaum: Hillsdale, NJ, 1981; pp 1–55.
53. Gunzelmann, G.; Anderson, J. R. *Cognit. Syst. Res.* **2003**, *4*, 57–76.
54. Wickelgren, W. A. *How to Solve Problems*; W. H. Freeman: San Francisco, CA, 1974; p 15.
55. Anderson, J. R. *Am. Psychol.* **1993**, *48*, 35–44, 36.
56. Mayer, R. E. *Thinking, Problem Solving, Cognition*, 2nd ed.; W. H. Freeman: New York, 1992; pp 180−181.
57. Matlin, M. W. *Cognition*, 4th ed.; Harcourt Brace: Orlando, FL, 1998; p 354.
58. Herron, J. D. *The Chemistry Classroom: Formulas for Successful Thinking*; American Chemical Society: Washington, DC, 1996; p 67.
59. Anderson, J. R. *Psychol. Rev.* **1987**, *94*, 192–210, 196.
60. Cardellini, L. *Come Risolvere i Problemi Chimici*; Ragni: Ancona, Italy, 1999; p 13.
61. Chi, M. T. H.; Feltovich, P. J.; Glaser, R. *Cognit. Sci.* **1981**, *5*, 121–152, 121–122.
62. Gabel, D. L.; Samuel, K. V.; Hunn, D. *J. Chem. Educ.* **1987**, *64*, 695–697.
63. Gabel, D. L. *J. Chem. Educ.* **1993**, *70*, 193–194.
64. Dávila, K.; Talanquer, V. *J. Chem. Educ.* **2010**, *87*, 97–101, 100.
65. Gabel, D. L.; Samuel, K. V.; Hunn, D. *J. Chem. Educ.* **1987**, *64*, 695–697, 697.
66. Genyea, J. *J. Chem. Educ.* **1983**, *60*, 478–482.
67. Friedel, A. W.; Maloney, D. P. *Sci. Educ.* **1992**, *76*, 65–78.
68. Waddling, R. E. L. *J. Chem. Educ.* **1988**, *65*, 260–262.
69. Waddling, R. E. L. *J. Chem. Educ.* **1983**, *60*, 230–232.
70. Lythcot, J. *J. Chem. Educ.* **1990**, *67*, 248–252.
71. Hoffmann, R. Private communication.
72. Watts, M. *The Science of Problem-Solving*; Cassell Educational Limited: London, 1991; p 15.
73. Silberman, R. G. *J. Chem. Educ.* **1981**, *58*, 1036.
74. Ashmore, A. D.; Frazer, M. J.; Casey, R. J. *J. Chem. Educ.* **1979**, *56*, 377–379.
75. Frazer, M. J.; Sleet, R. J. *Eur. J. Sci. Educ.* **1984**, *6*, 141–152, 151.
76. Cassels, J. R. T.; Johnstone, A. H. *J. Chem. Educ.* **1984**, *61*, 613–615.

77. Sumfleth, E. *Int. J. Sci. Educ.* **1988**, *10*, 45–60.
78. Byrne, M.; Johnstone, A. H.; Pope, A. *Sch. Sci. Rev.* **1994**, *75*, 103–107.
79. Bransford, J. D.; Sherwood, R. D.; Sturdevant, T. In *Teaching Thinking Skills: Theory and Practice*; Baron, J. B., Stemberg, R. J., Eds.; W. H. Freeman: New York, 1987; pp 162–181, p 167.
80. Simon, H. A. *Ann. Rev. Psychol.* **1979**, *30*, 363–396, 380.
81. Chandran, S.; Treagust, D. F.; Tobin, K. *J. Res. Sci. Teach.* **1987**, *24*, 145–160.
82. Staver, J. R.; Jacks, T. *J. Res. Sci. Teach.* **1988**, *25*, 763–775.
83. Ronning, R. R.; Mccurdy, D.; Ballinger, R. *J. Res. Sci. Teach.* **1984**, *21*, 71–82.
84. Witkin, H. A.; Moore, C. A.; Goodenough, D. R.; Cox, P. W. *Rev. Educ. Res.* **1977**, *47*, 1–64.
85. Anamuah-Mensah, J. *J. Res. Sci. Teach.* **1986**, *23*, 759–769.
86. Niaz, M. *J. Chem. Educ.* **1987**, *64*, 502–505.
87. Anamuah-Mensah, J.; Erickson, G.; Gaskell, J. *J. Res. Sci. Teach.* **1987**, *24*, 723–738.
88. Niaz, M. *Int. J. Sci. Educ.* **1988**, *10*, 231–238.
89. Niaz, M. *J. Res. Sci. Teach.* **1989**, *26*, 221–235.
90. Camacho, M.; Good, R. *J. Res. Sci. Teach.* **1989**, *26*, 251–272.
91. Padiglione, C.; Torracca, E. *Int. J. Sci. Educ.* **1990**, *12*, 187–194.
92. Atwater, M. M.; Alik, B. *J. Res. Sci. Teach.* **1990**, *27*, 157–172.
93. Johnstone, A. H.; Al-Naeme, F. F. *Int. J. Sci. Educ.* **1991**, *13*, 187–192.
94. Abraham, M. R.; Grzbowski, E. B.; Renner, J. W.; Marek, E. A. *J. Res. Sci. Teach.* **1992**, *29*, 105–120.
95. Niaz, M.; Robinson, W. R. *J. Sci. Educ. Technol.* **1993**, *2*, 407–416.
96. Rojas de Astudillo, L.; Niaz, M. *J. Sci. Educ. Technol.* **1996**, *5*, 131–140.
97. Noh, T.; Scharmann, L. C. *J. Res. Sci. Teach.* **1997**, *34*, 199–217.
98. Norman, O. *J. Res. Sci. Teach.* **1997**, *34*, 1067–1081.
99. Tsaparlis, G.; Kousathama, M.; Niaz, M. *Sci. Educ.* **1998**, *82*, 437–454.
100. Lee, K.-W. L.; Tang, W.-U.; Goh, N.-K.; Chia, L.-S. *Chem. Educ. Res. Pract. Eur.* **2001**, *2*, 285–301.
101. Simon, D. P.; Simon, H. A. In *Children's Thinking: What Develops?* Lawrence Erlbaum: Hillsdale, NJ, 1978; pp 325–348, p 345.
102. Larkin, J.; McDermott, J.; Simon, D. P.; Simon, H. A. *Science* **1980**, *208*, 1335–1342, 1339.
103. Priest, A. G.; Lindsay, R. O. *Brit. J. Psychol.* **1992**, *83*, 389–405.
104. Clement, J.; Lochhead, J.; Monk, G. S. *Am. Math. Month.* **1981**, *88*, 286–290.
105. Hegarty, M.; Mayer, R. E.; Monk, C. A. *J. Educ. Psychol.* **1995**, *87*, 18–32.
106. Johnson, D. W., Johnson, R. T., Smith, K. A. *Active Learning: Cooperation in the College Classroom*; Interaction Book: Edina, MN, 1991; pp 3:1–3:16.
107. Felder, R. M.; Brent, R. In *Active Learning: Models from the Analytical Sciences*; Mabrouk, P. A., Ed.; ACS Symposium Series 970, American Chemical Society: Washington, DC, 2007; Chapter 4, pp 34–53.

108. McKeachie, W. J. *Teaching Tips. Strategies, Research, and Theory for College and University Teachers*, 9th ed.; D. C. Heath and Co.: Lexington, MA, 1994; p 144.
109. Chiappetta, E. L.; Koballa, T. R. *Science Instruction in the Middle and Secondary Schools: Developing Fundamental Knowledge and Skills for Teaching*, 6th ed.; Pearson: Upper Saddle River, NJ, 2006; p 155.
110. Stice, J. E. *Eng. Educ.* **1987**, *77*, 291–296.
111. Ruiz-Primo, M. A.; Shavelson, R. J. *J. Res. Sci. Teach.* **1996**, *33*, 569–600, 569.
112. Novak, J. D. *Learning, Creating, and Using Knowledge: Concept Maps as Facilitative Tools in Schools and Corporations*; Routledge: New York, 2010.
113. Novak, J. D.; Gowin, D. B. *Learning How To Learn*; Cambridge University Press: New York, 1984; pp 36–37.
114. Herron, J. D.; Greenbowe, T. J. *J. Chem. Educ.* **1986**, *63*, 528–531.
115. Nakhleh, M. B. *J. Chem. Educ.* **1992**, *69*, 191–196.
116. Johnstone, A. H. *Chem. Educ. Res. Pract. Eur.* **2000**, *1*, 9–15.
117. Friedel, A. W.; Maloney, D. P. *J. Chem. Educ.* **1995**, *72*, 899–905.
118. Patrick, H.; Turner, J. C.; Meyer, D. K.; Midgley, C. *Teach. Coll. Rec.* **2003**, *105*, 1521–1558, 1521.
119. Lepper, M. R. *Cognit. Instruct.* **1988**, *5*, 289–309.
120. Renkl, A.; Mandl, H.; Gruber, H. *Educ. Psychol.* **1996**, *31*, 115–121, 119.
121. Cardellini, L. *Come Risolvere i Problemi Chimici*; Ragni: Ancona, Italy, 1999.
122. Cardellini, L. *Rev. Educ. Cien.* **2001**, *2*, 84–86.
123. Cardellini, L. *Found. Chem.* **2006**, *8*, 177–187, 182.
124. Fahmy, A. F. M.; Lagowski, J. J. *J. Chem. Educ.* **2003**, *80*, 1078–1083.
125. Lagowski, J. J. *J. Chem. Educ.* **2005**, *82*, 211.
126. Shulman, L. S. *Change* **2002**, *34*, 36–44.
127. Shulman, L. S. *Change* **1999**, *31*, 10–17.
128. Cardellini, L. *Int. J. Scholarship Teach. Learn.* **2009**, *3*, 1–6. http://academics.georgiasouthern.edu/ijsotl/v3n1/invited_essays/PDFs/InvitedEssay_Cardellini.pdf (accessed Apr 2010).
129. Cardellini, L. *Chem. Educ. Res. Pract.* **2006**, *7*, 131–140.
130. Cardellini, L. *J. Chem. Educ.* **2009**, *86*, 169.

Chapter 4

The Pursuit of Science in a Globalized Market: An Approach to Internationally Collaborative Science through Research Abroad Programs

Markita P. Landry*

Department of Chemistry and Center for the Physics of Living Cells, University of Illinois at Urbana-Champaign, Urbana, Illinois 61801
***landry2@illinois.edu**

The past decade has brought about a wave of globalization, which has affected everything from business to education. In the education sector, many programs have adapted to globalization by incorporating study-abroad components into their curricula. However, academic science in the United States has notoriously resisted the push to adopt a more internationalized approach to educating its students, and continues to do so to this day. Particularly at the graduate level, American students are reluctant to study abroad, which may have a drastic impact on the future of science in America. As a participant in the National Science Foundation's East Asia and Pacific Summer Institute fellowship program, I was able to experience many of the advantages a science graduate student stands to gain as they take their thesis to another country. Recognizing that a research abroad experience also comes with its risks and difficulties, much of the material presented here aims to facilitate and clarify the process of identifying and completing study abroad programs for science graduate students. These students, in turn, will be in a better position to help the field of science overcome many of the barriers currently hindering scientific progress.

If you are a graduate student in America, look around you. Chances are, you have become aware that nearly 40% of physical science graduate students in the

United States are international students (*1*). In certain fields, this proportion can be even larger. As a graduate student in Chemical Physics at the University of Illinois, I am well aware of the large foreign graduate student population here in the United States. Between the 8 members of our lab, we hold passports from 9 different countries, and speak over 10 different languages. However, I rarely thought about the effect studying abroad must have on graduate students. Furthermore, when so many of the world's graduate students opt to pursue graduate study outside of their home countries, why do American graduate students appear hesitant to do the same? One might think students would seize the opportunity to spend time in an exciting and perhaps even exotic foreign country. On the other hand, those same students are fully engaged in completing thesis projects and might perceive international graduate research experiences as "speed bumps" on their way to a graduate degree that may slow down their graduation timeline. Those students fail to appreciate that some of the program's greatest benefits are not immediately apparent.

One possible reason for graduate student hesitation towards studying abroad is that the high quality of graduate school in American universities makes it more attractive for foreign students to come here as opposed to American students studying abroad. While the United States is a global leader in many aspects of academic science, it is far from being the only scientific powerhouse in the world. For example, many nations such as China are emerging as future scientific frontrunners (*2*). Moreover, many areas of research are more easily performed in certain geographical locations due to local conditions that would be best studied on-site, such as research on geographically confined diseases (*3*). Lastly, and perhaps most importantly, research abroad provides insights into how other nations, cultures, and people approach scientific problems, allowing graduate students to be better prepared for jobs in a workforce that has become increasingly globalized. This begs the question: Why don't we see more American graduate students abroad? If not for an entire graduate degree, why not for a year? Or a semester? Or a summer?

One of the key reasons many American graduate students hesitate to go abroad must be due to the lack of information, particularly first hand anecdotal information, on the overall impact of studying abroad at the advanced education level. In an effort to fill that void, I will provide some feedback from my personal experience as an NSF East Asia and Pacific Summer Institute (EAPSI) summer research fellow at the University of Osaka in Japan. I will focus on how this experience helped shape my graduate career, and discuss tips on how to prepare for it so as to maximize your research productivity. I will discuss the *whys* of international research experiences for graduate students in the sciences, and how they have the potential to shape you and your career. Lastly, I will discuss how these international experiences benefit the scientific community as a whole by allowing you to become involved in breaking down barriers to cross-cultural exchange first-hand. Let this serve as a rough guide for how you, as a graduate student in the sciences, can become part of an increasingly large network of internationally driven science in today's globalized society.

A How-To Guide for the Aspiring Study Abroad Graduate Student

The National Science Foundation's EAPSI fellowship is a fantastic opportunity for American graduate students wishing to perform research abroad and establish long-lasting international collaborations. To apply, one is required to submit a research proposal for a project to be completed in collaboration with a laboratory in one of seven locations: China, Japan, Taiwan, Singapore, Korea, Australia, or New Zealand. To make the most of a graduate research exchange program, you will need to quickly adapt to a new culture, location, and language while performing research in a manner that will allow you to form long-lasting scientific collaborations. This is not an easy task. Much effort is required on your behalf to ensure that international research opportunities will yield optimal results, and that your time away from your home institution will be used to propel your thesis work and research interests. In an effort to help the aspiring graduate study abroad student, I will be candid in outlining the tasks I did (or wish I had done) in the preparation and execution of my research experience at Osaka University through the EAPSI program. This will help ensure a successful experience abroad while minimizing the probability of disrupting your graduate studies.

i. Identifying a Scientific Question

First and foremost, it is important to identify the scientific question one wishes to address in collaboration with a foreign laboratory. For many graduate students, applying for an international fellowship may be the first time the student independently writes a research proposal. Make sure that your proposed research project is directly relevant to your current research, and that the laboratory specializes in techniques and/or projects that will help propel your graduate career.

When drafting your research proposal, it is important to be realistic with regards to your research time frame. Many fellowships for international research such as EAPSI are short-term (a few weeks or months). It is important to consider this time frame as a fundamental limitation to your research proposal, and incorporate it blatantly in your application. However, in many cases it may be possible, and it would certainly be beneficial, to extend your research abroad into a long-lasting international collaboration between your home and host laboratories after your return. If this is a foreseeable consequence of your proposed research collaboration, stating it in your proposal will strengthen your overall application.

ii. Funding Your International Research Proposal

Once the scientific question has been identified, it is then necessary to find a funding source or a fellowship program to support the proposal. There are many funding sources for U.S. graduate students who wish to perform research abroad, varying in geographical location, funding amount, and program duration. The program in which I participated, NSF-funded EAPSI, ran from 8 to 10 weeks depending on the host country (*4*). Several other NSF-funded international fellowship programs exist, such as CESRI (Central Europe Summer Research Institute), and the Nordic Research Opportunity for NSF Graduate Research Fellows (*5*, *6*). For longer-term support, Fulbright fellowships are available for graduate students, post-docs, and other academics (*7*). Most of these fellowships require a scientific collaboration to be pre-established between the student and a host laboratory, and this must be explicitly stated in the fellowship application.

iii. Finding an International Host Laboratory

During the preparation of your application, you must concurrently identify potential collaborators and host laboratories that might help you realize your reasearch proposal. This potential for collaboration weighs heavily on the success of your application, and it is unfortunately one of the more difficult steps of the application process. In most cases, your soundest source of guidance for finding a host laboratory will be your graduate research advisor. Not only will your advisor have a better grasp of the intrinsic "feasibility" of your research proposal, but s/he will probably be better acquainted with other researchers in your field, and will be in a better position to suggest potential hosts.

It is also useful to consult with other lab members, or other scientists in your field. If time allows for it, prior to applying for international research fellowships, attend conferences in order to become more familiar with the work being done in your field. Moreover, the opportunity to speak with scientists about possibly establishing a collaboration is much easier to do face-to-face versus via e-mail.

You should also compose a short draft of your proposed research project and attach it to any correspondence that is sent out to potential hosts. In many cases, the host's primary language may not be English. While most science around the world is performed and published in English, individual researchers and even PIs (Primary Investigators) may not be as well-versed in the English language as their publications may suggest. For this reason, it is important to keep communications with the potential host laboratory as simple as possible, avoiding jargon and lengthy sentences while clearly communicating the goal of your potential visit. Once positive responses from potential hosts are received, they should be included in the fellowship application.

iv. Pre-Departure Preparation

Once funding has been awarded for your abroad research proposal, it then becomes necessary to prepare for your lab visit as thoroughly as possible prior to your departure. Prepare as much as you can before you leave, both scientifically and culturally.

If your host lab is located in a country whose language you do not speak, it is worth investing some time in learning key phrases you will need during your stay. Many countries, such as Japan, have higher standards for work etiquette than you will typically find in the United States (*8*). In my case, this made it imperative for me to know Japanese phrases that were necessary to show respect and a positive work attitude in lab. Every morning, saying *ohayo gozaimasu* (good morning, formally) was important. Since meals were often eaten with fellow lab members, knowing to say *itadakimasu* (I gratefully receive) before a meal and *gochisosama* (thank you for the meal) afterwards was also crucial. Being aware of cultural differences aside from language will also help minimize culture shock or gaffes on your part. Although cultural differences such as way of life, food, local traditions, etiquette and such are best learned on-site, it can't hurt to do a quick Google search for "[country name] traditions" in preparation for your program. It might have helped me avoid the awkward handshake/bow blend I ended up performing while meeting my host advisor for the first time.

Finding accommodation for the duration of your program can be difficult, particularly from across the globe. Personally, I underestimated this feat and it is something I wish I had done in a much timelier manner. While my accommodations ended up being spectacular, that was due more to serendipity than to careful planning. Most programs will provide assistance on finding accommodation for you, and some will have already arranged your accommodations. In other cases, your host laboratory will help you find housing. But in some cases, finding accommodation may be left entirely in your hands. If the latter is the case, I encourage you to communicate with your host laboratory and have them help you make these arrangements.

My case was a bit out of the ordinary. The international dorms where visiting students were typically housed were under renovation, and the only other option for on-campus housing was a fairly expensive hotel. I opted to take a non-conventional and slightly riskier route to finding housing: Craigslist.com. It is here that I found two young Japanese women looking for a third to share their quaint roof-top apartment in downtown Osaka. This option had many downsides: the commute to campus would be over 1 hour each way, and I had no way of checking the validity of this housing offer since it was uncertified. However, I decided to take a chance, and spent my summer with two absolutely wonderful Japanese women in the heart of the historic "tenjinbashisuji roku chome" district of downtown Osaka.

On a side note, if you do find housing on your own be sure are confident in your ability to locate your new home, especially if this is your first time in this new country. On my very first night in downtown Osaka, I struggled to find my new apartment. I thought I had been clever and cautious, printing out a detailed map of my new neighborhood with instructions on how to find my apartment. Building 4, apartment 1010. However, I found myself amidst row upon row of identical high rise apartment buildings. Furthermore, the numbering system was unlike anything I had seen elsewhere in the world: I later learned buildings were numbered according to when they had been built, not in ascending order along the street. There were several buildings marked 4 in the block. It was dark, and although Japan is renowned for its safety, I panicked as I dragged my two suitcases through mazes of buildings for what seemed like hours. Without a phone, and with only "professional laboratory" Japanese phrases with which to communicate, I felt desperate as I sat on the side of the street and tried to calm my nerves. Embarrassed, I stopped a passer-by to borrow a cell phone and call my roommate (whom I had yet to meet) by means of an elaborate game of charades. The feeling of relief as she came to fetch me on the street corner was unprecedented, and greeted me with a consoling hug rather than the traditional Japanese bow. The best way to avoid this? Kindly ask a member of your host laboratory (or roommates, if applicable) to meet you at your arrival port or station, and show you the way home.

You must also prepare "scientifically" for your abroad research. This type of preparation will be the most influential in determining the success of your research project. If you have not, by this point, thoroughly outlined your research plan for the summer, do so now. Most of this information should have been included in your research proposal, as well as the communications sent out to your potential host laboratories. But now is the time to break apart this research proposal into two main sections: what can be done in your home laboratory, and what must be done abroad in your host laboratory. Anything that can be done prior to your departure should be done now, since you will want to start on your summer project as soon as possible upon arrival in your host laboratory. In my case, this involved making 3 months' worth of my DNA substrates and shipping them to my host laboratory in Osaka prior to my departure. Now is also the time to communicate with your host laboratory about your personal expectations for scientific progress over the summer. In many cases, you may be assigned to work with another student or post-doc in the lab. It is fine to ask your host to suggest a person in the lab with whom to work. Keep a steady stream of communication with this lab member between now and your departure. Initially, send this person your research proposal, as well as any relevant literature pertaining to your project. In some cases, the fellowship program in which you participate may provide your host laboratory with a small amount of funding to cover laboratory expenses you may incur during your research experience. If this is the case, you should also communicate with your contact person about any laboratory equipment you may need for your project. If possible, request to have this equipment ordered prior to your arrival, so it will be at your disposal as soon as you arrive. In addition to communicating about research equipment, you should clarify whether you will need any training to use the laboratory equipment of your host laboratory. If so, you will need to factor this training time into your research plan. From your communications with your host person, try to gauge how much independence you will have in the lab and plan your work accordingly. Additionally, be sure to research and learn about any techniques and instruments you will need to use, but with which you are currently unfamiliar.

Lastly, ask program hosts (in the case of EAPSI, Japanese administrative contact personnel were made available via e-mail) if there are any particularities about the host work environment of which you should be aware. In the case of Japan, it was important for all career-oriented individuals, students included, to carry business cards on their person. Moreover, the manner in which these business cards were exchanged carried an extensive set of rules as well. For example, upon meeting a new colleague, the more "highly ranked" individual should be the first to offer their business card. Upon being presented with a business card, one had to study it carefully before putting it in their shirt pocket or purse. Putting the business card into pant pockets was seen as disrespectful, since it symbolized insolence towards the owner of the business card. The "lower ranking" individual then reciprocated, by offering their business card. Additionally, each time a business card was exchanged, both members involved in the interaction would bow, and business cards had to be presented with both hands holding the card. EAPSI program participants practiced this ritual several times during our training sessions, since it was a particularly important facet of the Japanese work environment. To be in a position where one was not able or willing to reciprocate the exchange of business cards is an enormous insult to the business card offeree. I was able to order business cards online, one side in English and the other in Japanese. I contacted students in my University's Japanese language program as well as fellow Japanese graduate students for help in translating my business card information.

Acquiring all of this information is time consuming, and often difficult, particularly if you and your host laboratory experience a language barrier. However, the time and effort it will save during your research experience is invaluable. Moreover, it will also help the members of your host laboratory prepare for your arrival, and will allow them to accommodate your project into the preexisting structure of their laboratory.

v. During the International Research Program

Now the fun begins! As soon as you step off the plane to your host country, I can guarantee that you will feel the rush of adventure, particularly if you have prepared well for your research experience. In my case, the first week after landing in Japan was spent in Sokendai, an academic campus south of Tokyo. It was here that the American EAPSI program participants met the other student participants from France, Canada, Great Britain, and Germany. Collectively, we became known as JSPS (Japan Society for the Promotion of Science) fellows, as our summer research program was co-sponsored by JSPS, a Japanese national science funding agency. We underwent a week-long "cultural acclimatization" orientation which included Japanese language crash courses, as well as a 2-night home-stay with a local Japanese family. Such on-site orientation sessions are common for abroad research programs, and are particularly helpful for students being hosted in countries with very different languages and/or cultures than one's own. The materials presented are typically tailored to preparing students for the professional laboratory environment of their host countries.

After this week-long orientation session, we each made our way to our host laboratories. I spent the latter half of my bus ride from Tokyo to Osaka practicing the handful of Japanese phrases I had learnt for the purpose of introducing myself to my hosts. My first day in the laboratory, I met my host PI and a few of the lab members. Amongst these 50 lab members were two other "gaijin" (foreigners), a rare sight in a scientific community heavily populated by, well, Japanese. In fact, "gaijin" presence was scarce in all of Osaka. On several occasions I noticed locals covertly taking pictures of me and my other "gaijin" friends with camera phones on the metro, a term we quickly dubbed as "gaijin-hunting". Funnily enough, though my host laboratory in Osaka spanned the entire building floor, I was assigned a desk right next to the only two other "gaijin", possibly as an effort on behalf of my hosts to make me feel more at ease.

On the first day, I felt hesitant to delve into my lab work immediately since I was a new and temporary student. I was surprised when nobody, including my "host contact person", helped me get started at the bench. While I did catch a few lab members' curious peeks towards my desk, my interaction with others remained minimal during the first few days in my new lab. I was pulled aside by a fellow "gaijin" lab mate who explained the cultural differences that might have made my frist day so awkward. First of all, as a Hispanic-American female chemical physicist, I created a very unusual presence in a Japanese physics laboratory. Not only are most Japanese scientists male (*9*), but non-Japanese scientists are few and far between (*10*). In fact, amongst the 50 or so scientists in my host laboratory, I was the only female graduate student, and one of only two foreign graduate students. However, the main cause of my discomfort stemmed from the inherent cultural differences for hosting new students. I learned that Japanese customs err on the side of independence when dealing with a new worker, and that I would have to initiate any help or training. In Japan, being proactive is seen as a trait that typically brings social discomfort, and is often seen as disrespectful in teaching scenarios.

For example, had my new senior lab mates given me a training schedule to learn to use the lab's instruments, this would have implied that I was deemed unknowledgeable or academically inferior to them. Instead, I was expected to seek advice from my peers as I encountered difficulties while independently adjusting to my new laboratory environment. Being unaware of these cultural customs, I refrained from seeking help from my new peers, worried I would be an unwelcome disruption to their everyday tasks. Meanwhile my Japanese hosts had done much of the same, out of respect for my scientific independence. Once made aware of this difference, I began rummaging around the lab for my materials, familiarized myself with the new instrumentation, all while asking my host lab mates for help whenever I encountered difficulties. I was spared the "learning experience" that my fellow "gaijin" lab mates underwent: One admitted to having wasted several weeks at the beginning of his abroad research experience waiting for his Japanese hosts to help him start a project. In the meantime, his Japanese hosts wondered why he didn't work, and why he did not come to them for help.

The point of this anecdote is to highlight how much influence cultural differences can have on your lab work, and on your overall research experience. While many westernized work cultures implement a "old teach young" approach for new hires, in Japan new students are treated the same as veterans, and it is up to the new student to seek help whenever (s)he encounters a problem. This can obviously create some productivity conflict if you are not aware of these unspoken cultural differences, and particularly if you are alone in encountering them. I would suggest for you to contact former participants of your abroad research program. The EAPSI program has a mandatory 3-day pre-departure orientation in Washington, D.C. that gives EAPSI participants the opportunity to meet EAPSI alumni a few months prior to embarking on their summer research experience. However, if your program does not offer such an opportunity, find some past participants of your program that are willing to share their personal experiences and the hardships they encountered for overcoming the cultural barriers in a foreign laboratory setting.

Throughout your research experience, but particularly during the first few days, observe the interactions between group members, and try to identify cultural differences in these interactions. The manner in which people interact, the laboratory tasks assigned to individuals, the level of formality and the style of dress (casual, business casual, etc.) of the lab members, work schedules, work expectations, and punctuality are but a few examples of what you should try to observe. Try to follow these cultural norms, not so as to assimilate to the host culture, rather out of respect for your hosts. It pays to be culturally aware of the differences in interpersonal interactions and scientific methodologies of your host country. While your host lab members will probably be lenient in not expecting you to know all of their cultural etiquette, they will very much appreciate you making an effort to identify and follow it.

As an American graduate student in the sciences, you are probably fairly accustomed to being in a nationally and culturally diverse school setting, as shown by the large percentage of international students at the graduate level (*1*). However, keep in mind that these same circumstances may not hold true for other nations such as Japan where foreign students are rare (*10*). They may be curious, inquisitive, and perhaps put somewhat on edge about having a foreign graduate student visitor, particularly if you are a female in a male-dominated field (*9*). On several occasions, I walked into a room where my male lab mates had been fraternizing, only to hear conversations being hushed immediately upon my arrival (ironically, as I did not understand Japanese, anyway). I later learned that certain topics of conversation (dating, personal lives, nightlife plans etc) are not "appropriate" to be discussed in the presence of female co-workers, and that doing so would be considered disrespectful. My Japanese lab mates, in these instances, had been trying to maintain their respect for me and the professional relationship that we had. Meanwhile, I had felt awkward and a bit left out during these occurrences. Similar situations occurred to a fellow EAPSI participant, based in China. She spent the entire 8 weeks of her program never having established communication with a lab mate with whom she shared a lab bench. Initially she assumed her attempts at conversing, and his blunt one-word replies, had been personal attacks against her character. However, she later learned that this lab mate was very self-conscious about his English speaking skills, was exceptionally intimidated by her presence, and was therefore reluctant to "shame himself" by attempting to speak English at a much lower level than her spoken English skills. It is of upmost importance to realize that you may be entering into a setting in which your new lab mates will not be accustomed to interacting with foreigners. Keeping a high level of cultural sensitivity during these times will put your lab mates at ease, and will facilitate the transition for them and for you.

One of the most effective ways of introducing yourself and your work to your laboratory is to give a short group meeting presentation on your personal and professional background. Ask your host advisor if this is feasible. Give a short PowerPoint presentation about your home academic institution, perhaps a few words on your personal interests, your research background, and what you hope to accomplish during your abroad research visit. "Putting it all out there" in this setting, early in your visit, will take the pressure away from future interactions with your lab mates. Furthermore, making your lab mates aware of your previous research experience may make them aware of any help you might need during your research visit, and might make them more willing to offer their expertise.

Once you feel you have established yourself in your lab, take some time to explore your new surroundings! For most EAPSI 2009 participants, the EAPSI program was the first time they had set foot in their host countries. Traveling within your host country is beneficial on two fronts: Firstly (and obviously), it gives you the opportunity to explore a new country. Secondly, it may give you the opportunity to meet other researchers and their labs. This can help increase your professional network within your host country and within your field. In fact, JSPS provided a substantial amount of money for JSPS summer research fellows to travel within Japan to meet other PIs and take tours of other labs. If you chose to do this (and I would strongly encourage you to do so), you can try to contact other labs within your host country prior to your departure, and ask if they would be willing to meet with you. Alternatively, you can ask your host research advisor if there are any local conferences that you may attend during your stay, or if (s)he has any colleagues in the field that (s)he feels you could benefit from meeting.

Lastly, just relax and make the most of this unique opportunity. Talk to other members of your graduate research exchange program (if applicable) if you get overwhelmed, and be sure to communicate any serious problems or concerns with program staff. And keep in mind, in the midst of your excitement while experiencing new foods, people, and places, that the worst of the culture shock typically occurs when (*11*)...

iv. You Return Home

This was certainly the case for me, and for many other EAPSI participants with whom I spoke after returning to the United States. Before embarking on my EAPSI adventure, I was told that reverse culture shock (i.e. the culture shock upon returning home) was much more severe than the culture shock of arriving in your host country. In my case, this statement could not have been more accurate.

For me, the process of re-acclimatization began with the tearful good-bye I shared with my two Japanese roommates at the base of Mount Fuji. It was also difficult to bid farewell to my host lab mates, who had helped me realize a significant part of my doctoral thesis. Culturally, there were also many aspects of Japanese culture I had grown to love, admire, and I had adopted as part of my everyday life. These ways of life also contributed to the reverse culture shock upon re-entry into the United States.

In Japan, respect for others, politeness, gratitude, and above all, integrity were traits held in the highest esteem for all: from the McDonald's cashiers who bowed, smiled, and profusely thanked every customer, to the Osaka University professors who instinctively bowed while talking to colleagues over the phone, I always felt respected and safe, never once felt threatened, mistreated, objectified, or endangered during my stay in Japan. In fact, when returning home early one morning after a night-long *tenjin matsuri* (annual local festival in Osaka), I spotted a man sleeping outside on a public park bench. He was dressed in his work clothing, and beside him were a neatly stacked pile of coins, bills, his mobile phone, and his keys. No doubt, he had removed these items from his pockets and had put them on the ground beside his bench, so as to sleep more comfortably. Nevertheless, it had never occurred to him that somebody would steal these items, because of Japan's low crime rates. Travel books constantly remark how astoundingly safe Japan tends to be: One particular travel book quotes the story of a tourist who forgot his camera at a popular Japanese tourist site. He returned the next day to find his camera at exactly the same location, untouched and unharmed. In my case, one of the most immediate elements of culture shock stemmed from how deeply I had become accustomed to this way of life. Upon returning home, I felt unusually repulsed by the gruff manner with which my bus driver demanded for exact change, by the apathy with which the librarian renewed my library books, I felt unusually paranoid walking home after dark, I mistook strangers' smile-less glances as forewarnings for aggression. I missed fresh sashimi, my Japanese friends and lab mates, I even missed being illiterate and being unable to understand the incoherent melody of the Japanese language in my surroundings. Gradually, I began to realize that these feelings were but mere reactions of my acclimatization to the Japanese culture. The bus driver wasn't being rude, the librarian wasn't apathetic, and nobody was out to get me during my walk home. Humans are beings of habit, and we respond strongly to changes in our environment. In fact, it isn't unusual for study abroad veterans to think more critically about their home cultures and societies upon their return home (*12*, *13*). This type of critical thinking, I would strongly argue, could help us all create better societies by adopting positive attributes from different cultures from around the world.

Lastly, upon returning to your home institution, you should make efforts to continue communication (and, even better, collaborative research) with your host institution. To this day, I occasionally participate in "language and science exchange" discussions with my Japanese lab mates over Skype. These hour-long discussion sessions are an exemplary medium to keep the lines of communication open between my home and host laboratory, and to continue the process of cross-cultural learning long after the exchange program has concluded.

I hope these tips and anecdotes will help you structure your research experience. It is no small feat, but the rewards that can potentially be reaped from a program such as EAPSI far outnumber the risks that plague the minds of many hesitant graduate students. With careful planning, an open mind, and a pocket translator, you can be ensured an academically and interpersonally successful abroad research experience.

Part 2: How Graduate Study Abroad Programs and International Experiences Lead to Increased Scientific Productivity

The United States is often seen as the global leader in all that is science, both at the academic and industrial fronts. International students contribute $12 billion annually to the American economy, making the US the leader in the field of international students of higher education (*14*). With America's stellar scientific reputation, and with its "dominance of the world higher-education market...", it is easy to understand why so few American students study abroad: 0.2 percent at the undergraduate level, and considerably less at the graduate level (*14*). Alarmingly, these rates are even lower, at all educational levels, for students of science (*15*). In becoming too comfortable at their home institutions, are American graduate students missing out on an essential part of graduate education?

Most of today's academic programs, from business to social work, are well aware of the effects globalization is having on their work forces. Many academic programs in these areas are beginning to incorporate mandatory international exchanges into their students' curricula, since students with international experience are likely to be considered better job candidates than their stay-at-home counterparts (*14*, *15*, *21*). Science being such an avant-garde field by nature, why are we lagging behind in the trend of academic internationalization?

When I was faced with the opportunity to participate in EAPSI, I was hesitant. I was hesitant to ask my PI if I could apply to EAPSI, I was hesitant to contact potential collaborators who might or might not have sufficient English proficiency to understand my emails, hesitant before clicking the "submit" button to my EAPSI application, I was hesitant when I accepted the award, hesitant throughout the pre-departure orientation, and hesitant as I boarded my 14-hour flight to Tokyo.

Primarily, I worried about my academic progress as I went abroad. It had taken me a solid year to get good lab results at my home institution, the University of Illinois. What made me think 10 weeks at Osaka University would be academically "profitable"? My decision to apply for an EAPSI fellowship was prompted by certain difficulties I had encountered during my Ph.D. thesis work. I had encountered two major hurdles, one stemming from instrumentation-induced photodamage to my samples (*19*), and the second stemming from observing unexpected behaviors of the biological system I study. I am interested in how certain proteins are able to identify a tiny DNA target sequence- usually only a few base pairs long- among thousands of base pairs of non-target DNA. Proteins are able to do so at surprisingly fast rates, and some, like the protelomerase protein I study, are able to do so without energy-rich cofactors such as ATP (*20*). This is analogous to finding your way to a new laboratory (like I had to do) in the middle of a very large foreign city (such as Osaka, Japan), in a car (without fuel, thankfully not my case) in a matter of minutes or seconds. However, I was having trouble extracting meaningful results from my data, and decided that I would need to approach my scientific questions in a different manner. I then was given the recommendation to apply for the EAPSI program, in order to take a different approach to solving my research hurdle. My goal was to image my protelomerase protein at a single molecule level, to visually observe its behavior during its interactions with DNA. Therefore, I contacted an expert in Total Internal Reflection Fluorescence Microscopy (TIRFM), Dr. Toshio Yanagida at Osaka University's graduate school for frontier biosciences, and asked him to be my EAPSI host researcher.

To say that my ten weeks in Osaka were a learning experience would be an understatement. New lab instruments, new culture, even a new circadian rhythm. Surprisingly enough, learning to operate TIRFM instruments was by far the easiest learning experience of the summer. First of all, I could not speak or read Japanese. This led me to make several mistakes only illiteracy could cause, such as eating my morning cereal with cream instead of milk, or accidentally washing my laundry in fabric softener instead of detergent. However, thanks to the astoundingly welcoming environment created by my Japanese lab mates and my Japanese roommates, I quickly became acclimatized to my summer living environment, and took advantage of this unique opportunity to be a graduate student in Japan. Somehow, in between my long days in lab, I had the opportunity to experience Japan's rich culture and history. I had the chance to visit each of my rommates' hometowns to celebrate Obon, a Buddhist holiday to honor one's deceased ancestors. I experienced the grueling 16-hour overnight hike to the summit of Mount Fuji. I got ambushed by the wild deer of Nara, and participated in the hectic annual city-wide festivals in Osaka and Kyoto. I took the high speed *Shinkansen* train to the volcanic cities in Japan's southern island of Kyushu. I even camped on the beach on Ikuchijima island with a handful of locals who spoke no English whatsoever, while attending a 2-day outdoor independent music festival not unlike a Japanese version of Woodstock.

These cultural experiences were fantastic, but the true tests of cultural aclimatization occurred in my host lab. In my Osaka University laboratory, I had to adapt rapidly to a new work environment, an environment dictated by a set of cultural laws with which I was unfamiliar. It was while recognizing and overcoming these cultural barriers that I learned the most useful cultural lesson of my summer as an international graduate student: Despite language and cultural barriers inherent in all cross-national exchanges, the drive for the pursuit of scientific discoveries is a universally upheld concept.

Recognizing cultural differences is instrumental for the success of any international research experience, but it is not an easy task. For the most part, a good approach to any new cultural environment is to play it safe and follow the lead. I observed interpersonal interactions between my new group members, and made mental notes of interactions that reflected Japanese culture. For example, it took but a few days for me to realize that interpersonal interactions in Japanese laboratories were much more formal than in the United States. Moreover, it became apparent that the laboratory had a rigid hierarchical structure, and that the level of formality with which one addressed another depended on the relative position of each person in this social hierarchy. For graduate students, deeper and longer bowing was necessary when greeting a post-doc or other senior level scientist. These observations were particularly important in group meeting settings, in which a lab member was presenting their work. In these cases, only group members considered "equal or higher" than the presenter in the laboratory hierarchy asked questions about the speaker's work. Even the laboratory directory at the entrance of the building listed lab members in order of their hierarchy within the laboratory. However, these social hierarchies actually made the laboratory more efficient and organized. Everyone knew their roles and tasks within the laboratory, and despite comprising over 50 members, everybody knew who to call on for help with experimental troubleshooting. Little time ever appeared to be wasted as everyone assumed their roles and their tasks from the moment they arrived in the laboratory, to the moment they stepped out. Much of this efficiency must have been due to the formal and hierarchical nature of the laboratory. It goes without saying that many laboratories across the world could benefit from taking pointers from Japanese laboratories, particularly in the realm of efficiency and cleanliness. If a seminar was to begin at eleven, *everybody* would be present at 10:55. If an instrument broke down, the person responsible for it would be quick to fix it. Everyone wore slippers in the lab, no shoes were worn past the laboratory entrance to keep it as clean as possible. And on Fridays at 10, without exception, all group members would abandon their current tasks, pick up vacuum cleaners, mops, and brooms, and clean the entire lab. I had never observed a more smoothly-running laboratory. It was more of a science factory than what I had come to know as a laboratory.

As time went on, I became keener to the differences in how Japanese labs function. I began to think more critically about some of the laboratory practices I had developed, and was able to merge my protocols with new ideas from my host laboratory; protocols that I still utilize to this day. Additionally, my experience abroad allowed me to "open doors" for my academic future. As a JSPS fellow, I was presented with an array of other opportunities to continue my studies in Japan after I will have completed my doctoral degree. JSPS offers very lucrative fellowships for post-doctoral researchers, as well as fellowships for professors who wish to spend some time teaching and researching in Japan. From what I was told by my host lab mates, the Japanese government has realized the need for internationalization of its scientific workforce, and has put fourth many incentives to diversify it. As such, I became aware of career opportunities outside of the United States. Moreover, I was able to establish a network of contacts within Japan, such as my host research advisor who invited me back to his lab for a post-doc. This type of networking would have been impossible to accomplish without being on-site, where my Japanese PI and colleagues were able to observe my work ethic, skill set, and productivity first hand. While reflecting upon all that I had personally gained during my time at Osaka University, I began to realize how dependant all of these learning experiences had been to me being away from my home country, out of my ¨comfort zone¨ so to speak. For this very reason, it struck me as odd that these types of international research experiences aren´t more ubiquitous amongst American graduate students.

While in Japan, it also became clear to me that in any academic laboratory setting, the science is only half the battle. Proper management of students and post-docs by the PI, and the manner in which scientific questions are addressed are crucial aspects of a well-established laboratory. The science half is universal: The speed of light and sound, the pull of gravity, the value of absolute zero temperature, the number of moles in a liter of water; all of that remains (to a reasonable approximation) the same whether you are studying these phenomena in Brazil or Sri Lanka. However, the *manner* in which these scientific questions are taught, learned, and researched varies greatly from laboratory to laboratory, and varies even more so from culture to culture. Teaching, learning, and researching are all socially structured and culturally driven approaches used to delve into the universal scientific questions we all wish to answer. Every country and culture approaches science in a different manner. And because there are strengths and weaknesses to all of these approaches, international exchanges allow students to sample these new approaches with open minds and bring home new ideas on how to advance science. I personally believe that the ability of members of the scientific community to 1- understand this and 2- work to use these differences productively through open minded collaborations, is directly proportional to the potential for the advancement of science on a global scale. One of the biggest challenges will be to encourage more American graduate students to pursue internationally collaborative research.

A mere 3 months at Osaka University was enough to convince me that there are certain learning experiences that simply cannot be acquired unless one ventures outside their home country. For one, students who spend a certain amount of time studying abroad will be exposed to that country's cultural environment, and will have to master the art of cultural tolerance and adaptation. These traits are crucial when dealing with a globalized workforce. But also, in dealing with cultural differences, students will naturally be introduced to the globalized nature of science first-hand. Students will find that while culture may dictate many aspects of their study abroad experience, science shares many universal traits. After all, science, unlike other markets such as business and politics, is not a product of an environment created by humans; rather science is inherent to nature thereby making it universal by default.

Learning to identify- and adapt to- cultural differences should be an integral part of every student's education. Research abroad experiences, particularly at the graduate level, provide students with an unparalleled medium in which to learn 21st century professionalism. Coming back to the idea of globalization, much emphasis is being placed on the ability of young professionals to interact with international members of their field in a professional manner. Such social and professional skills cannot be classroom- taught, and must be acquired (sooner rather than later) through interpersonal interactions. Markets have become increasingly responsive to the diminishing importance of national borders, and students within these markets must make the necessary adaptations to keep up with the changing times. I just hope that with all of America's stellar science, reputable graduate programs, and top- notch research institutions, American graduate students will not be left behind in the push for globalization to unify science on all national fronts.

Lowering Barriers to Cross-Cultural Exchange: Multi-Faceted Problems Require Versatile Solutions

Say you have a collection of 100 apples. Of these apples, 30 are fresh and crispy, 50 are mediocre, and the remaining 20 are old and rotten. If you are told to pick a harvest of 30 apples from this collection, without a doubt you would pick the 30 fresh apples every time. However, say you first take these 100 apples and randomly divide them into 10 groups of 10 apples. Once again, you are told to pick a harvest of 30, only this time you can only chose 3 apples from each group of 10. Statistically speaking, your harvest of 30 apples from the group of 100 will be better than the group of 30 selected by picking 3 apples from 10 groups. By imposing boundaries on the apples, you decrease the overall quality of your harvest, since some groups of 10 may have more than 3 perfect apples, and other groups of 10 may have no perfect apples for you to choose.

The same applies to science. Boundaries, whether they are national, political, socio-economic, racial, linguistic, or cultural, impose restrictions on the ability of the international scientific community to maximize its productivity. Now, some of these "boundaries", such as cultures or national borders, shouldn't be abolished in the name of science. Cultural differences, for example, provide the world with a necessary richness, and can even help advance science if approached properly. Personally, I enjoy taking part in new cultural traditions whenever I travel. I still remember my first trip outside of North America. I was 4 years old, and I went to Bolivia to visit my mother's family. As a 4-year old, I had a fairly limited understanding of culture. However, even from my eyes cultural differences became apparent, and I was in a constant state of amazement. Here, Christmas was smack in the middle of summer vacation. People lived mostly in villages or farms. People dressed differently; women wore colorful *cholita* dresses and often danced to traditional Bolivian music. The food was very different. When I returned home, I enthusiastically gave my kindergarten class a show-and-tell presentation on my experience. I told my fellow kindergartners about seeing wild alpaca, eating mango ice cream, and running myself nearly to exhaustion while playing a game of tag with my cousins 12,000 feet above sea level in the highest altitude capital city of the world, La Paz.

I felt 4 again when I landed in Tokyo. All of a sudden, not unlike my early years of life, I couldn't read, write, or speak (Japanese). I was amazed by the temples, the shrines, the women with kimonos, and the tea ceremonies. Centuries that shaped Japan as a country have allowed these cultural traditions to surface, a treat to experience for any tourist in Japan. And while some of these cultural differences, such as language, did pose difficulties for information exchange in my summer lab, it is these same cultural differences that make Japan such an interesting country to visit.

But cultures do much more than provide entertainment for tourists. Culture provides an overall structure for a country and influences everything from its laws to the religious views held by its citizens. Culture even shapes many ascpects of science. National and cultural dissimilarities need to be preserved and respected even if they might at times complicate the exchange of information. Therefore, in order to maximize the productivity of information exchange during- say- a summer research program in Japan, one must learn to work around these cultural differences in the lab while simultaneously learning to embrace them in every other aspect of their research abroad experience. Doing so is no easy task, but I found that making an effort to truly embrace a new culture during an international research experience allowed me to gain much more than new research techniques during my summer in Japan.

Culture must be preserved, yet it is the most prevalent dissimilarity that scientists will face during an international research experience. However, it can be used to the advantage of the scientific community if properly respected, learned, and even embraced. Over the course of my summer at Osaka University, I found that the extent to which my hosts were willing to accommodate me was directly proportional to the amount of effort I made towards learning and respecting Japanese cultural traditions. As a visitor in a foreign laboratory setting, your best bet is to keep an open mind about how a new culture will approach science. From eating moving octopus that was trying to escape from my dinner plate, to running my experiments using protocols I had never come across but were recommended by my new PI, I approached my international experience with an "I will try anything once" mentality. Though I never again ate a moving octopus, to this day I still implement laboratory techniques and protocols from my summer in Japan, techniques that I had never previously considered.

Furthermore, international exchanges automatically increase one's personal productivity as a scientist by amplifying the extensiveness of one's academic support network. By going abroad, I had the opportunity to share ideas with scientists outside of my immediate surroundings and I was able to gain a different perspective on how to approach my research. These experiences are doubtlessly beneficial to a graduate student's career, but cumulatively they increase scientific productivity on an international scale as well. Just as Charles Darwin noted that species evolve to acquire traits which allow them to best accommodate to their local environments, the manner in which science is performed has been optimized differently in each nation. Therefore, international research experiences allow members of one scientific community to experience alternate methods of approaching science. Subsequently, these new ideas are brought back to their home laboratories and become incorporated into their scientific repertoires, thereby increasing the overall productivity of the scientific community. Conversely, a scientist visiting a foreign laboratory will also bring in scientific tools and ideas from their home country, making international research exchanges rather symbiotic for both the home and host laboratories. In this manner, cultural barriers actually help science propagate, but only if scientists are willing to use these cultural differences to the benefit of science by actively participating in international research exchanges.

On the other hand, certain barriers to international exchange in the sciences are not advantageous. For example, barriers imposed on the scientific community by a country's economic or political standings can greatly stymie scientific progress.

Perhaps the most challenging, and therefore most frequently overlooked barrier, is the one formed by national borders and their corresponding politics. Political barriers are difficult to navigate, particularly because they are numerous and ever changing. Unfortunately tense political situations often dictate the degree to which scientific collaborations may be established between scientists of different nations. This leads to a lack of communication between scientists of these nations. Consequently, the extent of information sharing- which is crucial to the advancement of science- is compromised. Poor political influences can greatly limit the scope of scientific projects that are undertaken by scientists. We see examples of this everywhere. Venezuelan science has suffered greatly in the years of President Hugo Chavez, due to discrimination against or favoritism towards scientists based on political motivations (*21*). However, political scientific barriers are not strangers to top-tier research nations either, not even the United States. The stem cell funding ban imposed by President Bush in 2001 severely limited the amount and extent of research that could be done using embryonic stem cells, until the ban was lifted in President Obama's first term (*22*). Clearly, for an aspiring stem cell biologist, the United States might have been a questionable location to pursue such research interests in this time frame.

Even political policies that might be deemed irrelevant to science actually stymie the field quite markedly. Take the current hot-topic of U.S. immigration as an example. While immigration restrictions and firm visa requirements might be necessary to ensure the safety of the American people, they do more harm than good for American science. For many decades, the United States has relied on foreign-born scientists to help maintain its stellar scientific reputation and innovation output (*23*). Currently, foreign-born scientists contribute disproportionately to triadic patents originating from United States. Not only are there more patents originating from foreign-born scientists, but the quality of these patents are disproportionately superior and more likely to be commercialized. Consequently, foreign-born scientists in America contribute more than their share to American innovation both in scientific quantity and quality (*24*). Therefore, one would naturally assume that America would strongly encourage and even facilitate the immigration process for foreign scientists. However, the road to becoming a top scientist is not an easy one, particularly for foreign-born researchers wishing to contribute to science in America. While foreign-born researchers are clearly a lucrative asset to American science, many top foreign-born scientists face a tedious, year-long US visa application process (*23*). Political tactics put forth by the American government after the September 11th attacks, with the goal of making the United States a more secure country, are backfiring in the scientific community. The 2001 PATRIOT act, the Enhanced Border Security and Visa Entry Reform Act, and the Visa Mantis program have all contributed to a decrease in H-1B visas available for non-US born scientists who wish to bring their talents to the American scientific market (*24*, *25*). There is also a qualitative consequence to these political motions; they increase hostility of Americans towards immigrants, and foreign scientists in turn are less willing to immigrate to a country where they will face discrimination. In fact, trends in international student populations reflect changes in the political climate between the United States and other nations (*14*).

Disruption of studies is another main concern for foreign-born scientists who come to the United States seeking an undergraduate or graduate education. A student in my lab was stuck in Russia, his home country, for a solid three months waiting for his visa to clear. If American graduate students hesitate to participate in a 10-week international research programs from fear of disrupting their graduate studies, imagine the frustration of international students who become unwillingly cut-off from their American labs while stuck in the unnecessarily complicated process of renewing their visas. Perhaps the United States could take pointers from other countries; countries that face a similar dilemma of balancing their immigration laws to maximize scientific workforce flow without compromising national security. For example, another top recruiter of foreign talent, the United Kingdom, has instigated a "points" method that streamlines the visa process for talented workers before they even find a position in the U.K. (*26*). If the United States isn't quick to reform immigration policies, we may very soon lose the bulk of our scientific community- one that has allowed the United States to become the science powerhouse it is today.

Despite these political barriers, thankfully, we do still see an influx of foreign talent into the United States. However the efflux of American cross-cultural exchange remains stagnant at best. As a leader of the scientific community, the United States needs to push for internationalization of its academic curricula in the sciences, as it has for other areas of study. By refusing to step out of our national borders, American science is setting a poor example for both its students and also for international members of the scientific community. In doing so, we send a message that our science is self-sufficient, and that we are not willing to make efforts to step out of our home country to help science move forward. As a scientific leader, the United States is also left with many responsibilities, such as pushing for equality of international information sharing despite political tensions between countries. The United States should also do more to help developing nations become more active in science. Some steps to this end are being taken. Top scientific entities, such as Nature magazine, are working for the expansion of science to all corners of the world. Nature has recently launched several regional science websites such as Nature Middle East, Nature Asia, and Nature India (*27*). Facilitating access to scientific information will make scientists in these nations more willing and able to contribute to science-specific processes that need to be streamlined at the international level, processes such as peer-review.

Economic barriers are another unfortunate barrier to scientific international exchange. In certain cases, particularly for American scientists pursuing scientific experiences in countries that are not as developed as the United States, one must be sensitive to the confines imposed on scientific progress by economic limitations. Financially speaking, certain nations and their scientists may not be able to support research projects requiring expensive technologies. In this way, availability of funding for certain areas of research can severely limit the kinds of scientific questions that can be posed.

Economic limitations may also limit the level of scientific exposure students in these nations are able to receive. I recently gave a seminar on my research to students and faculty at *La Universidad Mayor de San Andrés* in La Paz, Bolivia. Despite being the top post-secondary institution in Bolivia, resources for students are relatively scarce. Above all, these students lack access to information on how to pursue scientific careers, and know relatively little about groundbreaking scientific advances being made in first tier research institutions. Despite these unfortunate differences, I believe that scientific communication with such nations is imperative. It is a responsibility which first-tier research institutions must undertake, lest we sacrifice the latent scientific contributions of their talented resident scientists.

Lastly, the most directly influential barrier to cross-cultural exchange in the sciences is the incongruity between science as a field of study, and science administration. Science is arguably the most universal field that exists today, being dictated by natural laws that do not vary across national and cultural borders. For that reason, it is also the field that would benefit most from internationalization. However, many of these difficulties originate from the science policy side of the table. Science education administrators need to eliminate the logistical hurdles science students face to study abroad.

As an undergraduate student, I completed two majors in the sciences: a Bachelor of Science in chemistry and a Bachelor of Arts in physics. My undergraduate institution, UNC Chapel Hill, offered an exchange program for science students called the Trans-Atlantic Science Student Exchange Program (TASSEP), in which I participated for a semester. UNC Chapel Hill is renowned for its undergraduate study abroad programs, boasting a 36.7% undergraduate study abroad rate. This is over 180 times the national average, making UNC the public university with the highest study abroad rate in the United States (*14*, *28*).Yet despite these encouraging statistics, I did not personally know a chemistry or physics student other than myself who studied abroad during their undergraduate years, while students in other fields studied abroad in herds. I encountered the same type of resistance while trying to encourage fellow graduate students in the sciences to consider international graduate research programs such as EAPSI.

What effect will these shocking discrepancies have on the future of science in America? Perhaps restructuring the field to include international exchanges, not unlike other fields have done, would go to great lengths in providing today's science graduate students with the international foundations they need to be maximally productive in this globalized society. At the level of science education, both at the undergraduate and graduate levels, restructuring academic tracks to include international exchanges (or at least facilitate them) would help incorporate elements of internationalization into students' curricula. Science as a field of study is unique in that it necessitates a "vertical" learning approach such that science courses *must* be taken in a certain sequence, cell biology preceding genetics, or differential equations preceding nonlinear dynamics and chaos. Other fields of study are more lenient in defaulting to a "horizontal" learning approach. Anthropology of indigenous societies need not precede sociology of Eastern European immigrants. Not having been exposed to the literary works of Henry David Thoreau won't impair your ability to study Shakespeare. This is an unfortunate fact of life for the internationalization of science education, because curricula for vertically learned subjects are consequently inflexible and therefore less amenable to the incorporation of study abroad components.

The incorporation of study abroad into science could be accomplished by working more closely with international post-secondary institutions to ensure that course credits students take abroad will transfer seamlessly onto their home institution transcripts. Another option would be to copy what other fields of study have done to ensure their students complete essential parts of their curricula in a timely manner, and gain experience learning and living in an international setting. For example, the University of Chicago's Booth school of business has expanded its campus to Barcelona, Spain, where it has a branch campus offering a University of Chicago MBA. This ensures that students are able to follow curricula set forth by the University of Chicago, and receive degrees from their home institutions while expanding their intercultural knowledge. Furthermore, schools in other nations are increasingly seeking American accreditation. Having American accreditation might help these schools to adopt curricula that are similar to those followed by American schools, and could facilitate international exchanges (*14*).

All of these barriers will be present, to varying extents, during an international research experience. For example, each of the seven locations offered by the EAPSI program varies in terms of political and economic standing. As a visiting student to some of these locations, one must be sensitive to the fact that certain nations may not be able to provide their scientific community with a level of funding to which you may be accustomed. Therefore, expecting these laboratories to provide you with certain equipment during your research experience may be unreasonable. Nevertheless, a positive research experience abroad depends much more on your personal motivation to succeed both in and out of lab, than on the types of instruments you are using. After all, the greater the dissimilarity between your home and host laboratory environments, the more rewarding the international experience will be as a learning tool for your career. Simply by taking the initiative to study abroad, despite all of the existing barriers to cross-cultural exchanges, you are already doing your part to help lower these barriers for future generations of scientists.

As American graduate students in the sciences, we need to realize that we becoming are part of a top-notch scientific community with an obligation to break down the barriers that are hindering universal scientific progress across national borders. Even as a graduate student there is much you can do, such as participating in international research programs like EAPSI, or taking the opportunity to visit local universities and science institutions during your international vacations. We must as a nation facilitate the process of international exchange, both for our own students of science and for foreign-born scientists wishing to study in America. We must amend our views and our actions as scientists in a manner that breaks down or works around political, cultural, socio-economic, and national barriers to scientific progress to create an environment in which science can thrive freely. Action must be taken, otherwise there will be no future for scientists that are being excluded from this globalized world, and the scientific community which knows no national boundaries will be deprived of their latent scientific contributions.

Final Remarks

It is evident that the barriers which continue to plague science are not problems that will be solved without collaborative effort from all members of the scientific community. Programs such as EAPSI are offered to encourage future generations of scientists to engage in collaborative research and open doors for scientific relationships that will strengthen ties between the United States and foreign scientific entities. By participating in such programs, you will gain both tangible (financial support, new professional networks) and intangible (personal growth, better world viewpoints) benefits that will help propel the future of science. Furthermore, professional international experiences will put you in a better position to help contribute to the solutions for the barriers to cross-cultural exchange. It will take efforts on behalf of all scientists to break down barriers that hinder scientific progress, and going abroad is an enormous first step. Going abroad will help you identify and actually encounter some of the barriers that are prevalent in science, which will make you more aware of how to approach science more effectively. There are also many other ways to approach these barriers that extend beyond the scope discussed in this chapter. You could try to bridge the gaps between elementary and higher education in the sciences by mentoring science-driven students through these transitive years. You can get more involved in the political workings of your community and push for the reform of policies that currently limit the advancement of science. You could even work with a choreographer to put together an interpretative dance of your research to help bridge gaps between science and the arts (*29*)! There is no better way to start your career as a scientist than by becoming involved in the amelioration of the field, particularly during your graduate studies.

The Japanese have a popular expression, "gambatte", which is known as "good luck" here in the United States. However, the meaning of "gambatte" is slightly more complex. It means to try one's best regardless of the result, encompassing the essence of Japanese culture and way of life. I encourage all scientists, from graduate students to fully tenured professors, to take it upon themselves to expand their cultural and scientific horizons through internationally collaborative science to the best of their abilities. After all, in the words of Mark Twain, "twenty years from now you will be more disappointed by the things you didn't do than by the ones you did do... Explore. Dream. Discover." 頑張って!

Acknowledgments

Special thanks to Toshio Yanagida, Komori Tomotaka, Thomas Martin, Lorenzo Marcucci, and Marc Landry. I would also like to acknowledge all members of the Yanagida and Chemla labs, participants and administrators of the EAPSI- JSPS 2009 program, and my roommates Mika Ishino and Kaori Arai for their help in realizing this study abroad experience.

This work is supported by the National Science Foundation, through the East Asia and Pacific Summer Institutes program under Award ID OISE-0913128 and by the National Science Foundation Graduate Research Fellowship under Grant No. 0913128. Any opinions, findings, and conclusions or recommendations expressed in this material are those of the author(s) and do not necessarily reflect the views of the National Science Foundation.

References

1. Oliver, J. *First-Time, Full-Time Graduate Student Enrollment in Science and Engineering Increases in 2006, Especially among Foreign Students*; InfoBrief NSF 08-302; Division of Science Resources Statistics (SRS), National Science Foundation: Arlington, VA, December 2007.
2. Adams, J. Get ready for China's domination of science. *New Sci.* January 6, **2010**.
3. Pastrana, S. J.; Clegg, M. T. *Science* **2008**, *322*, 345.
4. East Asia and Pacific Summer Institutes (EAPSI). National Science Foundation. http://www.nsfsi.org/. 2010.
5. Central Europe Summer Research Institute (CESRI). http://www.iie.org/en/Programs/CESRI. 2010.
6. Dear Colleague Letter: NSF Graduate Research Fellows Nordic Research Opportunity, 2008. National Science Foundation. http://www.nsf.gov/pubs/2009/nsf09011/nsf09011.jsp?govDel=USNSF_25.
7. Fullbright Programs. http://www.iie.org/en/Fulbright. 2009.

8. Huffman, J. L. *Mod. Lang. J.* **1989**, *73*, 95 .
9. Kuwahara, M. *Minerva* **2001**, *39*, 203.
10. Japan's tipping point. *Nature* **2009**, *460*, 151.
11. Sussman, N. M. *Int. J. Intercult. Rel.* **1986**, *10*, 235.
12. Carlson, J. S.; Widaman, K. F. *Int. J. Intercul. Rel.* **1988**, *12*, 1.
13. Williams, T. R. *J. Stud. Int. Educ.* **2005**, *9*, 356.
14. Altbach, P. G. *Change* **2004**, *36*.
15. Davis, T. M. *Open Doors: Report on International Education Exchange*; Koh Chin, H.-K., Ed.; Institute of International Education (IIE Books): Annapolis Junction, MD, 2004.
16. Festervand, T. A.; Tillery, K. R. *J. Educ. Bus.* **2001**, *77*, 106.
17. Rolls, C.; Inglis, A.; Kristy, S. *Contemp. Nurse* **1997**, *6*, 152.
18. Weinstein, S.; Barthalow, P.; Hamburg, M. V. *J. Sch. Health* **1976**, *46*, 599.
19. Landry, M. P.; McCall, P. M.; Qi, Z.; Chemla, Y. R. *Biophys. J.* **2009**, *97*, 2128.
20. Huang, W. M.; Joss, L.; Hsieh, T.; Casjens, S. *J. Mol. Biol.* **2004**, *337*, 77.
21. Sills, J. *Science* **2009**, *324*, 1514.
22. Wong, K. M. *Cell Stem Cell* **2009**, *4*, 111.
23. US visa nightmares. *Nature* **2009**, *461*, 12.
24. No, Y.; Walsh, J. P. *Nat. Biotechnol.* **2010**, *28*, 289.
25. Committee on Science, Engineering, and Public Policy (COSEPUP), Policy and Global Affairs (PGA). *Policy Implications of International Graduate Students and Postdoctoral Scholars in the United States*; National Academy of Sciences: Washington, DC, 2005. http://www.nap.edu/catalog.php?record_id=11289#toc.
26. America's got talent—Can it keep it? *Nat. Biotechnol.* **2010**, *28*, 181.
27. Science without borders. *Nat. Cell Biol.*, **2010**, *12*, 201.
28. Miles, R.; Spurr, K. W. UNC Leads Public Research Universities in Study Abroad Rate; Open Doors Report; Institute of International Education: New York, 2006.
29. Bohannon, J. *Science* **2009**, *324*, 332.

Chapter 5

There and Back Again: A Grad Student's Tale about Negotiating Studies on Two Continents in the Final Years of Graduate School

Charity Flener Lovitt*

University of North Texas, Department of Chemistry, 1508 W. Mulberry St., Denton, TX 76201
***chariteach@gmail.com**

During my 6 year PhD program, I studied abroad for nearly 15 months on two different fellowships; one short-term NSF funded summer fellowship to Central Europe and a year-long Fulbright fellowship to Germany. This chapter details the reasons I chose to pursue study abroad late in my graduate career, the steps I took to obtain and prepare for the fellowships, the experiences gained on the fellowship and the implications for my career. Comparisons are made between study at a research institute in Slovakia, a large private university in Germany, and a Tier 1 research university in the US. The final sections detail key elements of a successful study abroad which include: selectivity in choosing a host institution; setting clear research expectations for the time abroad; maintaining regular communication with the US mentor; befriending secretaries on both continents; and using freely available technology to remain in contact with friends and mentors on both continents. My experiences show that study abroad experiences can be successful late in the graduate career, provided that care is taken to maintain strong relationships on both continents.

Imagine a life where you are paid to work with people of different cultures. Where your lunchtime conversation takes place in three different languages. Imagine a career as a graduate student in which you have the opportunity to work simultaneously with top professors on two different continents where a lunch meeting over German pastries is followed by a video meeting with a professor in Illinois that is then followed by dinner in a 600 year-old student pub. While this story sounds more like a vacation than the life of a graduate student, this chapter discusses the factors that enabled me to spend nearly one-fifth of my graduate career working abroad. It will detail the difficulties associated with finding helpful mentors, securing funding, navigating personal relationships, and continuing graduate coursework at my US institution while 4000 miles away.

Oh the Places You'll Go: Experiences Abroad

As a graduate student, I had a wide range of research experiences. My graduate research spanned three distinct fields: synthetic organometallic chemistry, chemical education, and computational chemistry. As a result, I worked with four different research groups while in the US, three of them at the University of Illinois at Urbana Champaign (UIUC), and the fourth at an institution in Texas. I also received two fellowships to study abroad; the Central European Summer Research Institute (CESRI) Fellowship (*1*) which was funded by the US National Science Foundation and the Fulbright Student Fellowship which is funded by the US State Department (*2*). As a CESRI fellow, I worked with a small computational group at the Slovak Academy of Sciences (SAS) in Bratislava, Slovakia and a large computational group at Phillips Universität-Marburg (UniMarburg) in Marburg, Germany while the Fulbright fellowship allowed me to return to work with the UniMarburg group for a full year.

On both fellowships, I chose to live in university or research institute housing. This housing was relatively cheap and often extremely close to the building I worked in. In Slovakia, my building was on the Slovak Academy of Sciences campus, a mere two-minute walk to the building that house my office. In both German visits, I lived in the student village which was only two bus stops away from the science research complex on the top of Marburg hills. An added advantage to both buildings was that they were both close to major mass transit options. My office in Slovakia was three-blocks from a major bus stop while the student village was only five blocks from the main Marburg station.

Both the CESRI summer abroad and year-long Fulbright proved to be a mini-sabbatical experiences. Exposure to new research styles and groups renewed interest in my field and gave me different perspectives on my projects. My US research project, which had seemed like a dead end, suddenly had many new directions. The two months after my first summer were some of the most productive of my graduate career and lead to a presentation at a national ACS meeting (*3*). Furthermore, the research opened new pathways in other projects. I returned to UIUC in July 2009 where I quickly wrote my dissertation and graduated the following fall.

Study Abroad Myths

If someone had told me when I first started graduate school that nearly 20% of the time spent on my doctoral studies would be in central Europe instead of central Illinois, I would have laughed. It was impossible for a graduate student from the US to research abroad. Everyone *knows* that the US has the best schools. Furthermore, study abroad was expensive, and my research supervisor would never let me go because he wanted me in his laboratory. I call these the Three Myths of Study Abroad. I soon learned that these myths weren't exactly true.

The Myth of US Scientific Superiority

It is not a myth that great science is done in US schools, and that studies in science schools in the US are highly sought after and highly-ranked (*4*, *5*), but that doesn't mean that world-class research only happens in the US. As detailed in chemical industry articles, the percentage of the authors from non-US institutions publishing in top chemistry journals such as the Journal of the American Chemical Society is increasing (*6*, *7*). With the emergence of the European Union and a unified funding structure for scientists all over Europe through the European Science Foundation, expansive projects are funded from non-US coffers. High-quality research in many countries is performed by independent institutions that focus solely on research. For example, Germany produces phenomenal research from the famed Max Planck Institutes, and many of these institutes are affiliated with universities with students working jointly at the Institute and the university. In my research group in Slovakia, the graduate students at the Academy of Sciences were students at the university downtown. However, their published work only listed their affliliation as the Slovak Academy of Sciences. Interesting research can be found abroad if only because you are forced to leave the familiar university paradigm and interact with researchers in non-university locations.

The Myth of Expense

Fellowships for international graduate study in chemistry are relatively undersubscribed, making them widely available. The US State Department-sponsored Fulbright Program awards over 1500 fellowships each year to graduate students, and students in the sciences are encouraged to apply (*8*). Scientists can further increase their chances of obtaining one of these fellowships by securing a letter of invitation from the potential host professor. Such letters of invitation are relatively easy to obtain, as very few professors will turn down an opportunity to welcome a graduate student whose financing is already provided by an external entity. Outside of the Fulbright program, the National Science Foundation offers fellowships for graduate students (*9*) and welcomes grants with international components. Check the fellowships offered by the country with the research professor with whom you want to study. One of my American colleagues in Marburg was able to complete a Master's program in Germany using funds from the German Academic Exchange Service (*10*). The Hessen Academic Consortium was another resource available (*11*).

The Myth of Impracticality

The most daunting myth to a new graduate student is the myth that research professors in the US do not want to send their students abroad. However, the reverse of this myth is that few professors would turn down the opportunity to brag about students who earn international fellowships. Such fellowships, especially well-known ones like the Fulbright, come with prestige. Even short summer fellowships can lead to collaborations with professors from all over the world which can greatly help the research group through the infusion of new ideas and new collaborators. Research groups based in multiple countries have access to multiple funding sources.

If appeals to prestige and scientific opportunity fail to convince your supervisor, it is important to emphasize how the time abroad will allow you learn a new skill from the world's leading experts that will enhance the research in the group.. Bonus points are typically awarded if you can then teach the skill to other members. Finally, appeal to the professor's budget; each semester or quarter that the student's studies are supported by a fellowship is a semester that the professor does not have to pay for the student. These arguments worked in my favor in arranging summer fellowships, especially since the fellowship would enable me to learn a new skill that one else in the group knew.

While taking a summer abroad to study research ended up being one of the most positive changes in my academic career and helped my group produce work in a new field, the downside was that it greatly increased my desire to work abroad. As such, I couldn't wait to apply for a full-year fellowship. Thankfully, my graduate advisor was amenable to a full-year abroad since the short summer stint had been so fruitful. The full year that I spent at UniMarburg gave me the tools to finish my doctoral research. I was able to finish my degree less than 4 months after to returning to the US.

The Decision and Application Process for Short Term Study

I wish I could say that deciding to seek fellowships abroad came after a deliberative process where I carefully researched faculty at various institutions and determined the best possible location for research. In fact, the initial decision to study abroad came from a challenge by a fellow graduate student, who described to me his fully funded fellowship from the National Science Foundation to conduct research in Germany. I remember thinking, "Wow, I wish I could something adventurous like that, but my research advisor would never let me go, I have too much work to do here at Illinois." As an undergraduate I had spent nearly every summer abroad, but when I began graduate studies at the University of Illinois of Urbana-Champaign (UIUC) I thought that I needed to focus on my lab work. Travel would only decrease my research productivity and distract me from my research goals. Besides, the deadline was three days away; there was no way I could prepare an entire application in time.

However, another friend of mine was researching in Switzerland. I told him about the fellowship and he suggested that I apply anyway; the worst that could happen would be that the fellowship committee would say, "no." I asked my research advisor about the fellowship the next morning. In my discussion, I made sure to emphasize two things, obtaining mentorship from professors in my new field of theoretical chemistry and receiving funding the entire summer, so he wouldn't have to pay for my summer stipend. Since the research group (and the university) was in a tough financial situation and my group did not have the expertise to help me understand some of the theoretical chemistry, my boss was more than happy to help me with the paperwork.

I spent the next two days preparing the 8-page application, which required a two-page research proposal. Before I could prepare the proposal, I had to select a mentor. I utilized two methods, relying on my advisor's expertise and searching the web. The fellowship, the Central European Summer Research Institute (CESRI), was funded by the National Science Foundation and organized by the Institute of International Education. CESRI restricted research interactions to six countries; Germany, Austria, Poland, Hungary, and the Czech and Slovak Republics (*1*). My graduate research advisor had no collaborators in those countries but we had a collaborator in Texas who knew a lot of people in theoretical chemistry. He suggested three names, but strongly recommended Prof. Gernot Frenking in Marburg, Germany. I cross-checked the names he gave me with research papers and determined that Prof. Frenking would be a great fit. His work in bonding analysis could help explain some reactivity trends we saw in a class of molecules known for dihydrogen activation. However, I needed to provide secondary selections in case the first selection could not be accommodated. For the second selection I relied entirely on internet and SciFinder search results, using keywords such as density functional theory + *keyword for one of the 3 projects I was working on + country I want to visit.* I searched conference proceedings for Central European countries and reviewed research papers for over a dozen scientists. I ultimately selected a professor in Bratislava, Slovakia who specialized in creating highly accurate models of magnetic molecules. My research group in the United States had recently synthesized a simple molecule with very interesting magnetic properties so my back-up project would be the study the magnetic properties this molecule in Bratislava, Slovakia.

The packet was finished in time and mailed via overnight shipping. Communication with the fellowship staff later revealed that my packet arrived only a few hours before decisions were made. Fortunately, I was accepted to the program in mid-April. My situation was unique. I was one of the few applicants who listed a possible collaboration with a Slovak scientist so the CERSI administrators provided funding to send me to Slovakia and Germany for the two halves of my fellowship, beginning in mid-May 2008, so I had one month to finalize details for both projects, arrange housing, and arrange transportation between all countries. In subsequent communcations with my mentors, I had to be mindful of the 7-hour time difference between central Illinois and central Europe. In a two-week period over 60 emails were sent between IIE staff, my collaborators, and myself.

Three transatlantic phone calls later, we finally established dates for my visits to both research groups. Fortunately, my German host would be out of the country for the first three weeks of my fellowship period and my Slovak hosts would be unavailable for the last four weeks of the exchange period, so splitting my fellowship between the two groups was an ideal solution. I would spend three weeks in Slovakia and four weeks in Germany.

Deciding and Applying for a Long Term Fellowship

The decision to apply for a Fulbright Fellowship for September 2008-July 2009 occurred on a much more relaxed timeframe, albeit much faster than the five months recommended by the Fulbright commission. After returning from Germany in late July, I realized how much I enjoyed working with the group in Germany. My summer as a CESRI fellow was fast-paced, but provided really interesting ideas for research; ideas that were developed into a presentation at a national meeting a few months after I returned from Germany. I wanted to return to Germany to further develop my project and the relationships I had started.

I visited the UIUC Fellowship Office to discuss a fellowship for the next year. The advisor was optimistic, especially when I mentioned that I already had an established relationship with a research group in Germany. He then listed several different types of fellowships in Germany, including the Fulbright and a German Academic Exchange (DAAD) fellowships (*10*). I decided to apply for the Fulbright since it guaranteed funding for a full year and had the earliest deadline.

The Fulbright application process involved much more work than the CESRI application; including a faculty interview, a language test, and multiple levels of screening. The Fulbright website strongly recommended that potential applicants start working on their application a minimum of four months before the deadline, while I only had one month. My research host in Marburg, Germany extended an invitation to continue research in his group for a full year, which became a large component of my final Fulbright application. Since I already knew where, and for whom I wanted to work, I thought the most difficult part of the process was finished. I assumed that I could reuse my CESRI proposal and personal statement for the Fulbright proposal. However, the fellowship advisor suggested that my research proposal had to be completely reworked so the research description could be understood by a chemistry novice, but evaluated by a chemistry expert and fit into the two-page limit. This meant that terms such as catalysis had to be explained succinctly but accurately without using scientific jargon. A document that I thought would take four days to write took nearly four weeks. The two-page document was reviewed by 15 people of diverse academic backgrounds, including my undergraduate chemistry professor who had received a Fulbright five years earlier, my research advisor, a former CESRI fellow in mechanical engineering who was on a Fulbright fellowship in Rome, and my sister who was a first-year year graduate student in divinity school. The proposal had to be understood and approved by all in order for my application to succeed.

I submitted the application through the university in early September. At the beginning of October I was called for an interview with a faculty member in chemistry, a German language professor, and a few other professors who previously had an international study abroad fellowship. This committee reviewed my application and then assigned a confidential score, which was forwarded to the national Fulbright Commission. The committee also provided recommendations for improving my final application.

This final application was submitted to the national Fulbright Commission late October. To be accepted to the Fulbright, an application has to pass through the US committee, and then be forwarded to the committee representing the proposed country in which the research is to take place. In early Febuary, nearly three-months after I submitted my application, I learned that the US national committee had recommended favorably my application to the German committee. Over half of the applications to the German committee were normally accepted so my chances were good. Two months later, in April, over seven-months after submitting my first application, I learned that the German committee had extended a full-year fellowship to me and placed me into a language tutorial. The language tutorial started in mid-August so I had three months to pack up my life for a year and move to Germany (*12*).

Preparing for the Transition Abroad

The busy weeks of preparations before departure for both CESRI and Fulbright fellowship were similar. In both cases, I had to secure waivers from my US university to ensure that my student status would not change despite my absence. I also had to facilitate communication between my US and European advisors to ensure that the projects ran smoothly. Since my work was computational in nature, I did not have to ship samples of reagents ahead of time but I did have to set-up computer accounts on servers in both countries to ensure that I could start research as soon as I joined the groups. I also needed to ensure that my laptop had the proper European power supplies and that I had phone numbers, addresses, and email addresses of contact people in each country.

Both fellowships required an initial financial investment. In the case of the CESRI fellowship, I had to book my own tickets to and from Europe and for travel between my fellowship site and the CESRI orientation in Budapest. Since my CESRI fellowship took place in many countries, I decided to purchase a Eurorail pass, which proved to be a wise investment as the initial $300 cost gave me over $700 worth of rail travel. My travel expenses were be paid for by the fellowship but I did not receive my first fellowship check until after I departed from the US. This meant that I had to use my graduate student credit card to book all flights. I posted so many airline and train tickets to my credit card that the company called to confirm that the purchases were not fraudulent. In the case of the Fulbright fellowship, I was able to book my ticket using a special student travel service that billed the ticket to the Fulbright program. My primary financial outlay consisted of shipping costs to mail essential books to Germany.

Preparation for the Fulbright took much longer than the short-term CESRI fellowship. For the CESRI fellowship, I simply left my apartment for two months. I continued to pay rent but my belongings were secured by my housemate. For the Fulbright, I had to move my belongings out of my graduate apartment into long-term storage. My parents and friends provided space for free but I spent a substantial amount of time packing and moving belongings. I also had to obtain explicit leave of absence from the graduate school at Illinois and apply for admittance to Phillips-Universität. Futhermore, I had to provide a research timeline to my thesis committee to prove that my year-long absence at Illinois woud result in timely graduation. This last step was extremely important as I was five years into my degree program and the committee had expressed concerns that my progress would be impeded by further travel.

One futher complication to the Fulbright was the need for travel papers. The CESRI program lasted less than three months so I could travel on a US passport without a visa. For the year-long Fulbright I had to obtain a student visa so I had to make sure that I packed my high school and college diplomas, should the university require these. Since the German Fulbright program was well established I could rely on the expertise of the German *Kommission* to help smooth over the visa process. The director of the language program personally carried my passports to the German registration office and returned them two days later with the required student visa. However, one of the scholars that year almost did not complete his visa requirements on time and came close to being kicked out of the country. He had the presence of mind to show his original flight ticket into Germany, which allowed him to complete registration and obtain the visa.

Distinctions Between Research in the US and Other Countries

Every research group is different, and differences between groups are often greater than differences between different countries. However, there are some key frameworks to remember when preparing to work in different research and national cultures.

First, the Language Barrier Is Basically Non-Existent in the Sciences.

Although chemistry is a language unto itself, English is widely spoken. Research chemists must have a working knowledge of English in order to read most chemical literature, so many take chemistry classes in English. Even if a course is taught in the student's native language, at least one of the books for the course will most likely be in English. For example, while in Germany I sat through two courses on Quantum Mechanics. The lectures were given in German but two of the four suggested books for the course were in English. Additionally, one of the German books was also a German translation of an English-language text.

In the research environment, a working knowledge of English is usually necessary to participate in multinational groups. In the research group in Marburg, I worked with scientists from 5 different continents and the Middle East. Love of the research field often transcends love of country. We all had similar interests in chemistry, and any cultural differences served only as fodder for lunchtime discussions. Although we sometimes had difficulty explaining our favorite foods, we were always able to discuss our research, either through words or drawings, which is typical for chemistry research. The only time I struggled to communicate with my research group was during my time in Slovakia, when confusion arose between my Slovak hosts, who were trained as physicists, over terms used in electron paramagnetic spectroscopy. Interestingly, this confusion did not arise from misunderstandings of the English language, but rather the nomenclature barrier between physicists and chemists when discussing electron excitations.

However, there is a caveat. Despite talk about the prevalence of English in science publications, I found the experience of learning German for chemistry to be particularly enlightening for my research. For example, I gained a much greater pictorial understanding of the English word "approximation." In my mind the word, always associated with a mathematical guess, gained greater meaning when it was paired with its German counterpart, *Näherung*. I learned this word after already learning the phrase "In der Nähe" which means, "in the neighborhood" or "nearing". With a greater appreciation for *die Näherungen* (approximations) used in chemistry, I began to use them more often in my research, which greatly increased productivity (*13*).

A Second Thing That I Learned Is That How One Spends or Saves Time is Culturally Dependent.

In the US, people are constantly overscheduled. In order to accomplish all necessary tasks, they 'save' time to write grant proposals or read journal articles by extracting time from other activities. Thus many US professors (and grad students) must be skilled in multi-tasking. It is a rare sight to visit a conference lecture without seeing a dozen laptops filled with people checking emails or proofreading papers during the speakers talk. Time "saved" by checking email in lecture is then used to accomplish other tasks. This approach is effective for accomplishing large numbers of tasks in a short period of time, but can be mentally taxing on the person (*14*).

In contrast to the US, time in Germany and Slovakia was spent carefully. Multi-tasking activites such as checking email during a lecture were frowned upon. I once attempted to check email during the course of a lecture in a large (400+ person) conference, but decided against this proposition when my typing raised a few eyebrows from other people seated in my row. Bus also schedules enforce a strict working time; work had to be accomplished in the time allotted for it. Working past this time was discouraged through lack of transit options, official registration for late night entry, and at times even by the introduction of vicious dogs into the area surrounding the research buildings in Slovakia (*15*). This strict separation, enforced by dogs in Slovakia and locked doors in Germany, meant that deliberate effort was made to spend time focused on research alone. Phones were viewed as distractions and often ignored during this time. However, since science does not operate on a 9 to 5 schedule, this sometimes meant being separated from the office when work needed to be done. In Germany, I was able to use internet at home to make-up for lost laboratory time while in Slovakia I had to remember to print articles to read later, or just learn to finish work on time.

Another way time was spent was another cultural difference between working in Europe and working in the US; I was actively encouraged to travel on weekends. The travel was meant to refresh my mind so I could work efficiently during the week. In Slovakia, my hosts were ethnic Russians who migrated to Germany two days before the Berlin Wall fell, which led to many long and enlightening conversations. My hosts took me on many hikes around the countryside and made sure to explain the cultural significance of each one. On one weekend, I was invited to join a hiking trip through the Lower Tatras mountains where we ate a picnic lunch in the courtyard of a castle destroyed by Napoleon's army. During a Sunday afternoon walk, my hosts walked me through the former demilitarized zone where the Soviet military shot anyone who tried to cross the 20-meter Danube River to Austria. We explored an abandoned World War II bunker and even found Roman coins on the side of the road. In Germany, I used my student travel pass to visit many UNESCO world heritage sites in Germany; at one point I visited 5 sites in under 24 hours. My birthday was spent in a September snowstorm on a glacier in the Austrian Alps, throwing snowballs at German colleagues. The month before Christmas was filled with visits to *Glühwein* (hot spiced wine) stands in German Christmas markets. I discovered the joys of efficient train travel on weekends, as I traveled to visit Fulbrighters in other German cities or former German exchange students in Berlin. On weekends not on the train, I hiked the hills around Marburg and surrounding Hessen, which were the inspiration for the collection of fairytales now known as the Stories of the Brother's Grimm. In the spring and summer I traveled to conferences in Austria and rural Germany. The ability to formally disconnect from my research, whether because I was forced out by working conditions or because I wanted to travel allowed me to more easily focus on the time I wanted to *spend* on research.

Third, I Learned the Difference between Working in an Established Department versus a Dynamic Department.

The two universities I interacted with most, UIUC and UniMarburg are well-established locations for chemistry. The Chemistry department at UIUC has been around for over 100 years while the chemistry program at UniMarburg has existed for nearly 600 years and was the first university in the world to establish a professorship of chemistry (*16*). Although both schools produced phenomenal chemists, in Germany, divisions between subfields were great, which restricted collaboration between groups. However, while at UIUC, I worked in three research groups for three different subfields of chemistry (inorganic, organometallic, and chemical education) without needing departmental approval to do so. During my six years at UIUC, the school added four new interdisciplinary fields. Although the current German government has sought to decrease the divisions, it was difficult to break more than 500 years of tradition. In Slovakia, old divisions between groups remained from the Cold War Era, but new initiatives and a need to share limited resources helped to increase collaboration in the newly revamped science curriculum. Consideration of how these traditional divisions can impact your scientific productivity is extremely important when dealing with researchers from other countries or preparing to study abroad.

Making the Most of Your Time Abroad

Traveling abroad in the middle of a degree program can be very difficult, especially if the degree requires exact knowledge of a specific topic. These tips helped me to craft a successful program:

- Be selective;
- Set clear goals;
- Be receptive and open-minded;
- Communicate regularly and often;
- Make friends with the secretaries/administrators;
- Get a good laptop.

Some of these tips (especially the last three) may seem obvious to anyone who has traveled abroad but these six tips helped make both fellowships fruitful and exciting.

Be Selective

It is important to carefully select a professor and location that will work well with your research. Just as picking the correct research advisor is important for a successful graduate career, picking a host for study abroad is critical. The host should be knowledgeable in a topic that you want to study, and should also have time available to meet with you, or he or she should have a group that is large enough to assist you. It is also helpful if the group has hosted guests before. In my research group in Slovakia, I was the first guest of the group, so the hosts were unsure of what to do with me (they did an excellent job for the short time I was in town). In Germany, the group hosted visitors on a nearly monthly basis, so they were accustomed to providing quick access to the tools that I needed and even provided social outlets for a short-term visitor.

Ideally you should meet the mentor before you apply to work with him/her but as a graduate student, your options for travel are more limited. In my experiences, I was unable to meet with the mentors beforehand but my advisor in UniMarburg was recommended by the professor I worked with in Texas because the professor in Texas had worked with the UniMarburg professor before. The UniMarburg professor was not recommended purely because he produced great research, but because he was known to work well with people from all over the world. My interactions with the professor in Texas had been excellent so I trusted his recommendation, which ended up working well for me.

Set Clear Goals and Expectations

It is also important to set clear goals with both your research mentor in the US and the host mentor in the guest country. If you applied for a fellowship, your research proposal should list these goals. However, it helps to write down specific items that you want to accomplish in your visit. Above all, be realistic – everything takes longer than you think it will. For short-term projects, establish possible research outcomes BEFORE boarding the international flight. For example, for my first short term (four-week) project in Germany, I emailed the question I sought to answer for my project. They responded by emailing journal articles that related to my project. When I arrived, the professor had already assigned a post-doc to help train me and my computer was already set up with the group. We spent one hour discussing my research and pointing out how their work would complement the results I had already obtained. Before I left the US, I had already transferred important files to the group computer server so since my computer was already set up, I started running calculations that first day (*17*). The post-doc then checked my progress on a daily basis and we always compared the results of my calculations to the question I had asked.

For long-term visits (e.g. a Fulbright Fellowship), set one major goal for the year and multiple secondary goals that you can use to check your progress. For example, my proposed Fulbright research proposed using a bonding model to explain differences in reactivity for a series of metal complexes that could activate methane. However, after two months in Marburg, I realized that I had not progressed on this goal because I was having difficulty running the computer program that performed the calculations. So I set a secondary goal of learning the skills for this computer program that I was unfamiliar with but this research group used regularly. I developed a side project that was simple to execute but required me to master use of this code. Thus, instead of becoming dejected over my inability to complete my primary project, I was enthusiastic to see the progress on the secondary goal which helped keep my focused on my original goal. After completing the goal and mastering the code, I was able to progress on my primary task.

Also, it is extremely important to stay aware of any deadlines at your home institution. Being proactive is a good way to eliminate confusion before it begins. For example, I left for my Fulbright the year that I had to submit an original research proposal to my thesis committee in the US, an essential step toward finishing all of the requirements for my doctoral studies. This goal was articulated to my host professor before I arrived in Germany, and he allowed me "time-off" from my regular research to prepare this proposal. In addition, before I left UIUC, I emailed all of my thesis committee members with a list of the goals for my time abroad in order to mitigate confusion on whether I "worked efficiently" while away. These goals included attending lectures in quantum mechanics and making clear progress on my thesis, which I recorded by emailing progress reports to my thesis advisor on a monthly basis. When I returned to my US institution one year later, my committee knew the goals that I had completed and did not object to my graduation three months later.

Be Receptive and Open-Minded but Not Easily Offended

This advice can be applied equally to anyone traveling in a foreign country or to scientists in diverse fields (*18*). Read and learn as much about the country as possible before you travel but realize that situations will arise that the book can't address. I read up on European etiquitte before I left which helped me both keep from offending my labmates and from taking offense at unexpected behaviours (*19*).

The study abroad experience forces you to break your normal patterns and is useful in that it forces us to think critically about things we ignore in our own culture. It's easy to discount cultural differencs as bad instead of just simply different. I listed one example above in how the European labs closed earlier than I was used to in the US. When I first learned that I had to leave the lab early, I became upset and declared that science could not be constrained to a 7am to 9pm box, it required 24-7 dedication. However, as I discussed above, I realized that the fact that labs closed early allowed me time to decompress from my work. I didn't stop thinking about my science, I just stopped focusing on the little problems in computational code and reflected on the big picture of my project goals. Oftentimes, the solution to the little problems became clear once I returned the next day. Futhermore, I realized that my labmates were not wasting time by avoiding the office for the weekend, but instead they were spending time on other projects, like walking the trails of Marburg.

Of course, as a computational chemist, I was able to continue my research outside of the the normal work hours as long as I found internet access. However, for a researcher performing a 17 hour reaction, limited access to the lab would have been difficult. In this case, the US model would have been more helpful. By being receptive and open-minded, I was able to craft a schedule that helped me spend and save time as I needed.

A caveat to this tip is, *"don't be afraid to ask questions, instead be wary if you find yourself not asking any questions."* I still found it difficult to complete simple tasks such as buying a stamp from a machine or navigating the transit system. In these situations, I simply learned to ask questions from the people around me. Even if we could not speak the language, I learned how to utilized hand guestures and mime to express my point. However, its important to read the etiquiette guides because in some countries, simple behaviours such as biting your thumb can be taken as insults (*19*).

Regular Communication

While abroad, you must communicate regularly with both the host professor and the research committee in the US. Specify exact times for these meetings with your professor in the US, especially when time zone changes are involved. Meetings with my professor at UIUC took place every week over Skype web chats, which were free for both the professor and me. Since this time was on the calendar each week, I actually met with my US research advisor more often from Germany than I would if I worked down the hall from his office. Scheduled meetings encouraged accountability and helped me keep to my research goals.

I highly recommend that any student studying abroad obtain an Internet connection account that offers international calling through the computer. Skype is the most widely used, and it is possible to purchase a local US number that can be used by friend in the US to call your computer, no matter where you are in the world. Other programs include Google Voice, AOL instant messenger, and Yahoo chat. With my US number through Skype, family, friends, and colleagues could call at any time using their US phone number to reach me in Germany. The phone call would be charged at US rates even though I answered in Germany.

Befriend Departmental Secretaries

Secretaries can navigate paperwork for the host institution, including processing paperwork that you really should have provided in person, and they can help you navigate through the beaurocracy in any country. The secretaries at UIUC made sure that I submitted all graduation paperwork in advance, while the secretary at UniMarburg provided translation help for student fees and even helped me figure out how to pay a confusing German doctor's bill. The secretaries can help schedule appointments for a professor that claims he is too busy to meet with you and easily navigate multiple university beaurocracies.

Purchase a Good Laptop Computer

You should purchase a laptop that is portable and easy to use. I recommend a laptop with an internal webcam and microphone, as it simplifies your ability to contact people all over the world. I was able to use my webcam to present at group meetings to my research group in the US while I finished work in Germany. Furthermore, my US professor stayed logged into Skype during the day, so I could easily call him when I had a brief question, much like briefly stopping by his office while on the way for coffee. The internal webcam also helped reduce the emotional distance between family and friends on different continents since I could see them when we talked.

Following these tips can help make a study abroad proceed smoothly. Time and money saved with these tips will allow you spend more time discovering your host country and culture.

Is It Worth It? Pros and Cons To Study Abroad during Graduate School

In making the decision to study abroad, there are a few pros and cons to remember before making the final choice..

One of the biggest disadvantages was separation from close friends and family for a full year, which resulted in many nights feeling lonely and isolated, a common occurance for students on these trips (*20*). These feelings were lessened somewhat by living in a student dormitory where I could talk to other students. Time in Slovakia was short and the research group took me in as family, so I often had dinner plans 4 nights out of the week. However, the SAS apartments lacked an Internet connection, which meant that I was cut off from communication from 11PM to 6AM every day. Although 7 hours seems to be a short time, it can be very long for a night owl like me, especially with the knowledge that my friends and family in the US were awake. I survived by reading books. With hindsight, I should have spent more time communicating with other students in the building, which would have kept me distracted on lonely nights.

During the first few months of my Fulbright, before I gained proficiency in German, I found it very difficult to communicate with my housemates and thus felt very lonely. However my German dorm had Internet access, so I could communicate with family and friends at night (which was daytime for them due to the seven hour time change). My youngest sister was in graduate school at the time so we spent 1-2 hours 'studying together' over webcam. Video conferencing also allowed me to attend dinner with my family. As I gained German proficiency, I started spending less time at the virtual kitchen table in the US and more time in the dormitory kitchen, where I befriended my German, Polish, Turkish, and Danish housemates. Through these housemates I learned about Eurovision, German politics, German opinions of US politics and the best time to buy bread at the bakery. Our kitchen had the advantage of a balcony that overlooked the castle of Marburg, so we spent many nights barbequing brats and laughing at German and English jokes.

Although I gained many friendships in Europe, spending a year abroad required me to suspend my life in the US for a year. Both of my sisters got married during my year abroad. I was able to fly back for their weddings, but I was unable to attend the wedding showers. We tried to video chat for some of the party games but it was not the same as attending the showers in-person. Additionally, I became engaged less than three months before I left for Germany. My fiancé (now husband) lived on the West Coast of the US, which meant that we had to juggle schedules that were nine time zones apart, greatly restricting our ability to communicate. One of the least favorite questions during this year was, "What are the plans for your wedding?" Living 6000 miles away from my fiancé and 4000 miles away from potential wedding venues meant that planning for married life could not take place until I returned to the US.

Additionally, many of my close friends graduated before I returned from Germany, which meant that the graduate student support network that I had used to survive the first five years of graduate school had to be rebuilt once I returned to the US. Since thesis writing consumed most of my free time in the US, I did not have time to rebuild these relationships, which increased the amount of isolation I felt. Although I was able to maintain strong communication channels with my US professors with weekly Skype meetings, I was not able to keep up with my US lab mates. The lack of interaction did not affect my research because my project goals did not overlap with my lab mates very much. However, upon returning to the US to finish thesis writing, my group had undergone many changes. In addition to changing office spaces, the composition of the group had changed since many old members had left and new members had joined so I almost felt like I was joining a new group. Although my new lab mates were supportive during the brief time I was back in the US, I greatly missed the camaraderie that used to sustain me through the rough times in graduate school.

Other negatives included the time to prepare the applications for the study abroad, complete all the university paperwork, locate alternate housing arrangements, and take language courses. The move between countries took about four months away from graduate research, which slowed down my graduation timeline. However, my lab mates claim that the year abroad sped up my rate of graduation because I worked more efficiently in Germany and Slovakia than at

UIUC. In those places, I knew that my time was limited, so I had to work quickly in order to accomplish my research goals in the allotted time frame; increased work efficiency helped to negate the time spent preparing to leave the country. Overall, despite my year abroad and five months spent in transit or preparation, I graduated only five months later than other people in my class.

I also missed conference interactions in the US, as neither my graduate mentor in the US nor my host professor had any incentive to pay for round-trip international airfare. I considered using my personal funds to attend at least one conference in the US but I decided to use my time abroad to focus on scientific networks outside the US. (I used the saved finances to travel around Europe.) At UniMarburg, I attended three scientific conferences, all of them specializing in a different aspect of physical chemistry. At these conferences, I discovered the advantages of being the only native English-speaking student. My presence as the American student on a prestigious fellowship meant that I could gain easier access than other students to some of the American speakers at the conference. At one conference in particular, it turned out that the only other American had once worked with one of my advisors in the US, so I was able to chat with her on the way to some of the conference activities. Also, when American professors visited UniMarburg, I was often invited to dinners with them because I was the only American student in my field.

Furthermore, because I was engrossed in a new research and conference environment, I was exposed to new ideas and new ways of thinking. This exposure forced me to think differently about many aspects of my research project - including simple changes such as electronic data collection, and as a result I became a more dynamic thinker.

Because I was unable to visit conferences in the US, I was unable to network with US colleagues. However, by being outside the US, I was able to gain an outsider's view of the employment situation and my time abroad proved interesting to recruiters in the US. All other things equal, many chemical companies will hire a candidate with international experience over one without, especially if that experience involved working in a research environment (*21*). Although searching for a job in the US while living abroad meant that in-person interviews were prohibitively expensive, I was able to participate in several phone and webcam interviews via Skype. In one interview with a large industrial company in the US, the recruiter asked extensively about my interactions with colleagues abroad. I was eventually offered a position in the US even before I left Germany, which was four months before I defended my PhD work. A demonstrated ability to work with people of different cultures was cited as a key reason why the position was offered to me. Thus, I can say that the interaction with colleagues in two different continents proved to be a greater boon than detriment to my career.

Concluding Remarks

Overall, my time abroad more has been more of an advance than a setback for my scientific career. As a result of studying abroad in Germany and Slovakia, I was able to gain expertise in diverse subjects while studying on fellowships that brought prestige to my career. My work proceeded relatively distraction-free because I was on limited timelines that accelerated the pace of my graduate research. I was exposed to a network of scientists and ideas outside my standard expertise while still able to accomplish research goals. Travel abroad is useful in that it forces us to think critically about things we take for granted in our own culture. Even simple differences like a small cup of coffee in the US being larger than a large cup of coffee in Germany puts us in a mindset that enables us to reevaluate long hold traditions. These experiences then force us to become more cognizant of how we approach our work and our teaching. For example, "Is my approach to using written exams effective for enforcing learning, or is the German model of oral examinations worthy of introduction into the classroom?"

On the cultural level, I learned a great deal about a different part of the world. As a child growing up during the end of the cold war, I was brought up with the image that former communist block countries lacked social structure and rich cultural histories. My experiences in CESRI countries proved this image false. I found that each country possessed a very unique cultural imprint. In Hungary, I discovered how great Hungarian minds were behind advancements in theoretical chemistry and I now see how many math professors are Hungarian immigrants. In Poland, I learned that the culture of a one great nation that gave us Nicolas Copernicus and Madame Curie. In the Czech Republic, Slovakia, and Austria I saw remnants of the great Hapsburg domain along with scars from a century of wars. When I wanted to take a break, many fascinating locations were just a short train ride away.

There is one incredibly large 'con' to studying abroad - once you start, it's difficult to stop. After a summer in Slovakia, I could not wait to visit another research lab abroad. After returning from a yearlong study in Germany, I cannot wait to see what other labs I can go work in. To paraphrase Bilbo Baggins, the adventurous hobbit from Lord of the Rings, "It's a dangerous thing, stepping into a new environment, once you start down that reaction path, you don't know where your chemistry might take you."

Acknowledgments

I am grateful for the opportunity to experience study abroad with and for the support provided by the Institute of International Education and the German-American Fulbright Commission. I also wish to thank Drs. Charles and Teresa Spicer, Prof. Paul Kelter, and Andrew Lovitt for advice and suggestions. I am also indebted to my graduate mentors, Prof. Greg Girolami and Thom Dunning, for allowing me to leave the group for a year to study abroad.

References

1. Central European Research Institute (CESRI). Institute of International Education. http://www.iie.org/programs/CESRI (accessed April 2, 2010).
2. The Fulbright Program. Institute of International Education. http://www.iie.org/Fulbright (accessed April 2, 2010).

3. Flener, C.; Frenking, G.; Girolami, G. *Inductive Effects of Cyclopentadienyl Substituents on Osmium Dihydrogen Bonds*, 235th American Chemical Society National Meeting, New Orleans, LA, April 6–10, 2008; INOR 187, Poster.
4. Academica Rankings of World Universities, 2009. Shanghai Ranking Consultancy. http://www.arwu.org/ (accessed April 2, 2010).
5. G-Factor International Ranking 2006. http://universitymetrics.com/gfactor2006top300 (accessed April 2, 2010).
6. Which Countries Publish the Most Cited Scientific Papers? Thomson Reuters Newsletters. http://science.thomsonreuters.com/news/2007-06/8392239/ (accessed April 2, 2010).
7. *The Future of US Chemistry Research: Benchmarks and Challenges*. The National Research Council's Committee on Benchmarking the Research Competiveness of the United States in Chemistry, National Academies Press: Washington, DC, 2007; ISBN 978-0-309-10533-0.
8. Competition Statistics for US Student Fulbright Program. Fulbright, Institute of International Education. http://us.fulbrightonline.org/thinking_competition.html.
9. Office of International Science and Engineering. National Science Foundation. http://www.nsf.gov/div/index.jsp?div=OISE (accessed April 2, 2010).
10. German Academic Exchange Service. http://www.daad.org.
11. Study in Hessen. http://www.study-in-hessen.de/front_content.php?idcat=23 (accessed April 2, 2010).
12. NOTE: I should note briefly that although the Fulbright application includes a language test, I was not required to speak fluent German unless language fluency was key to the success of my research proposal. Since I wanted to work in a group in which English was the primary language of communication, I did not need German fluency in order to complete my project. However, since the Fulbright fellowship was founded in part as a diplomatic outreach effort, the ability to communicate with local people is an asset. Prior to my Fulbright, I could communicate in German on a basic level, but I was not conversational. The German Fulbright offered an opportunity to take a six-week language tutorial before the official start of the fellowship. After the tutorial and a year of living in a German dorm, I often found myself talking for hours *auf Deutsch* and navigating German life with ease, something I found to be one of the most exciting aspects of my time abroad.
13. For examples of the impact of approximation in Science see (a) Bartlett, T. The Gospel of Well-Educated Guessing. In *Chronicle of Higher Education*. http://chronicle.com/article/The-Gospel-of-Well-Educated/65351/ (accessed July 9, 2010). (b) Swartz, C. *Back-of-the-Envelope Physics*; Johns Hopkins University Press: Baltimore, MD, 2003.
14. Multitasking—Bad for the brain? *Science* **2009**, *325*, 1053.
15. Note: During my time in Slovakia, I lived at the Slovak Academy of Sciences, an area of Bratislava surrounded by fences and barbed wire. In order to keep people from breaking into the Science buildings to steal

expensive equipment, dogs were released onto the academy grounds after 11pm each night. These dogs were incredibly vicious and had even been known to attack people in cars. I had to stop my research at 10:50 PM each night so that I could travel safely from the chemistry building to my academy apartment. If I did not leave at this time, I would be forced to spend the night in the chemistry building because the building securities officer would not let me leave.

16. Everts, S. The dawn of academic chemistry. *Chem. Eng. News* **2009**, *87*(44).
17. Note about preparation – experimentalists may have to ship chemicals and samples to the site ahead of time, while computationalists download molecular geometries to computer clusters.
18. Storti, C. *The Art of Crossing Cultures*; Intercultural Press: Yarmouth, ME, 1989.
19. Morrison, T.; Conaway, W. A. *Kiss, Bow, Shake Hands: Europe, How To Do Business in 25 European Countries*; Adams Media Cooporation: Avon, MA, 2007.
20. Preparation materials for the Fulbright list strategies for coping with loneliness and depression, suggesting that this is a common problem with students who study abroad.
21. Companies have mentioned this in personal interviews, but this issue has also been discussed during several ACS Careers talks. To view these talks, visit www.acs.com/careers.

Chapter 6

Research Abroad: Coming to a Decision, Making It Worthwhile, and How It Has Impacted My Career

LeighAnn S. Jordan*

Department of Chemistry, Michigan State University, East Lansing, Michigan 48824
***leighannjordan87@yahoo.com**

Undergraduate research is key to a successful career in chemistry. In performing it, you have two fundamental choices. You could conduct the research in the country of your undergraduate institution and stay relatively within the confines of what you know. Alternatively, you could go on an unforgettable adventure and travel abroad in order to do research. Personally, I can say that the experiences I had during my research in a foreign country fundamentally changed both my personality and the way I see the world for the better. They have shaped who I am today, where I am in my chemistry career, and who I want to become. There are numerous research abroad programs available to undergraduates but choosing the right one can be difficult. This chapter first discusses how to decide upon a research program and then provides suggestions on how to make your trip worthwhile. The final part of this chapter presents information of how my research abroad has affected my career, and some possibilities of how it might affect yours.

Coming to a Decision

Research is key to a successful career in chemistry. It allows you to explore theories, expand your knowledge and test your own ideas. Following my sophomore year in college, I participated in a summer research program in a state different from where I grew up. However, it was still in the United States, and relatively within my comfort zone. The techniques I learned and knowledge I gained made the summer worthwhile, but my interest to study abroad was truly sparked when I became aware of how international the science community really is. After interacting with the international postdoctoral fellows and speaking with my friends who had traveled abroad, I was curious to see what options were available to study chemistry abroad. There were many meetings concerning summer and semester research abroad programs, but I found that most were set up for humanities and language studies. In order to find those programs aimed towards chemistry majors, I had to embark on a different search method: Google. I then learned that there were many international research programs available for the sciences, specifically for chemistry majors, and started my analysis of my study abroad options.

The first step in the decision process is to decide when to go abroad. Chemistry programs are highly regimented such that course work must be completed in a very specific manner. Deviations from this order might cause for changes in class scheduling and possible delays in degree completion. It is important to find out if your school has had other chemistry students study abroad and when they studied abroad. The school might have a template for people to follow in order to graduate in a timely manner. It would be much easier to follow a plan than to make your own and try to convince your institution that your plan would fit into an undergraduate schedule. Knowing if your credits from another university transfer and if your school allows time in its curriculum in order to study abroad would be useful. Due to specific lab and class requirements, I would have needed to stay an additional semester if I studied abroad during the fall or spring semester. Be sure to discuss your plans with your advisor and understand the flexibility of your specific curriculum.

An additional consideration when deciding when to study abroad is the possible setbacks associated with each time option; each semester will have its pros and cons. For me, a summer abroad meant missing another entire summer of family and friends. It was also a summer that I had planned to work at my summer job to replenish my bank account for my senior year. However, in my case, the ability to graduate on time was more important to me. This ruled out traveling during the fall and spring semesters, therefore I refined my search to just those programs held between the spring and fall semester.

To my surprise, an online search for summer research abroad opportunities turned up multiple options. Spain, France, Australia, Italy, United Kingdom, Japan, Germany, and China were just a few of the countries that appeared during my initial search. It seemed that if I wanted, I could find a program in any country to study for the summer. With so many possibilities before me, I realized that it was necessary to decide what I wanted to take away from this experience before I continued my search. I had to decide what was more important, the funding, the country I traveled to, or the research project I would be working on. I decided that my goal was to immerse myself in the culture and have a positive research experience. Your personal goals may differ, but be sure to take the time and define them before you choose your country and program. The program you choose will be determined by your goals. Each country has a unique culture. I wanted to experience all of them, which did not help to narrow down the choices. So I had to focus my search something more than just the cultural immersion and be more detailed in what I meant by a "positive research experience."

I wanted my research to be as pleasant and productive as possible. However, I realized that in order to learn as much as I could during the summer, I would need to be able to ask questions, receive answers, and in general, communicate well with the people around me. My biggest fear was going to another country and not being able to communicate inside and outside of the laboratory. Therefore, I limited my search to those options which offered language programs or where the primary "in-lab language" was English. The ability to communicate would allow me to learn more efficiently both inside and outside the laboratory.

The next key thing to consider when choosing a program was the provided funding. Many programs did not show the amount of the stipend on their website and/or you needed to be accepted into the program before you would be informed of the amount of your stipend. Other websites gave approximant stipend amounts. The stipends ranged from absolutely nothing to a few thousand dollars. I did not come across many programs that covered travel, housing and other expenses. Some programs had access to discounted travel fares and could place you in an inexpensive living arrangement. Money may make the world go around, but you do not want to limit yourself to a program that pays for everything; you might be limited to one choice. Know your financial situation and the limits of what you can spend during the summer. I knew I would not be able to completely pay for a summer or semester abroad on my own. My search was thereby limited to those programs that offered a stipend. The stipend had to be significant to at least pay for the flight to and from my country of choice.

As any scientist would do, upon narrowing my search, I started to analyze my options. I created a spreadsheet about each of the countries, the programs available, their funding and if language classes were available. I wanted to push the limits of my comfort level, so I applied to programs in countries that were not necessarily English based. I had studied Spanish for many years and was interested in the Spanish culture. The applications were separate for each of the universities in Spain and most stated that the lab language was Spanish. I also found a summer research program in Germany that provided dozens of research project choices and mentioned that the lab language was English. I had never studied German but they offered a two week intensive language program that was meant to prepare anyone to live in Germany and be able to communicate. The process of applying for the program in Germany was more convenient, as I only had to complete one application. I could then choose my top three research groups and send them individual personal statements. There were so many research options in Germany that I also had to narrow my decision by the area of research I was interested in. I felt my interest in the research was more important than the international ranking of the university where the research was being conducted; as many other students I encountered had ranked their choices. I also wanted to study something I was moderately familiar with but using different techniques.

After completing my spreadsheet, I ended up with a sizable list of research projects that I would enjoy dedicating an entire summer researching and could even see myself continue researching through graduate school. At this point, I inspected more details about these research groups that were available online. Some of the information that proved helpful was the size of the university/group, rankings of the university and other information. Personally, neither the size of the group nor the specific university mattered. I knew that I would not become lost at a large university because I would be confined to the chemistry department and buildings. A small group would not limit my contact with others; I could explore outside of my laboratory group.

Since the size and university affiliation of the research group did not condense my search, the geographical location came into play. I then began to research the city in which the research took place so that I could better rank my choices. Two clear options appeared; large city and small town. My undergraduate career had been in a small town, located within a square mile area. I could break out of my comfort zone and attend a university in a large city, exploring the city life. Or, I could refrain from facing complete culture shock by choosing a university in a small town. I realized that I could travel and experience the "city-life" without being engulfed in it and not quite so overwhelmed. For this reason, I ranked the universities in smaller towns first.

When, where and what are decisions that need to be made on an individual basis. "When does it best fit in your career, where you think you will enjoy your abroad experience the most, and what research interests you?" are three of the most important things to consider when deciding on participating in a research abroad program. Once you have chosen your programs and you have finished your applications, the only thing to do is wait to find out if you were accepted. The only limitation to the number of programs you apply to is your motivation.

Once I was accepted into the program in Germany, I booked my flight within a week. The sooner you book your flight, the more money you could possibly save. I then determined how much money I would be able to make during the year following my study abroad. I took this into consideration when making a budget for my summer. Other preparations for the summer consisted of reading my lab's recently published papers and emailing the Ph.D. candidate I would be studying under, who I still refer to as "my Ph.D." My Ph.D. and I discussed, through email, who each other was, what I could expect from the area where I would be living, and what it is like in the laboratory. I was not familiar with German etiquette, so I made sure that was part of the culture I learned. I was not sure how helpful the language program was going to be so I bought an English-German dictionary, phrase book and language starter program. I would recommend a translation dictionary and a phrase book to anyone traveling abroad, even if he/she has taken a few language classes. Before I was able to say auf wiedersehen, I was off to Germany for the adventure of a lifetime.

Making It Worthwhile

While in Germany, I maintained a journal (something I recommend you do as well as it's the best souvenir you can bring back with you). I have gone through the different journal entries that I wrote during my stay in Germany, and have compiled a list of tidbits you should consider in order to make your trip as worthwhile as possible.

The first thing when you go abroad is to not cut off communication with everyone from home. There will be so many different adventures, why not tell someone about them? It is not necessary to purchase an international cell phone to talk with your family and friends though; they are expensive. With that being said, an in-country cell phone is useful to keep in contact with your laboratory and other people you meet on your trip. You can buy a relatively inexpensive phone upon arrival to the country where you will study and purchase minutes as you go. For those who have any chance of being homesick, know that you can not prevent it but you can decrease the severity of it. Internet services such as Skype, Facebook and emailing were helpful tools to keep in contact with those people who are important to you back home. Since there are drastic time differences between Europe and the United States, the hardest part when you plan to have live conversations is trying to sync times. Take time to keep in touch with those family and friends at home. It not only helps relieve homesickness, it is a great way to help facilitate record keeping of what you have experienced.

Remember that memories are not perfect and unless your experiences are documented, many will likely be forgotten. Relaying your stories to friends and family back home, keeping a journal and taking photos are ways that will help you remember your time abroad. As I look at my journal from my summer abroad, I am still surprised of how many specific instances that I do not remember, even after less than two years. I initially kept a daily journal that soon turned into a weekly journal, about my experiences inside and outside of lab. Luckily, one of my hobbies is photography and it was second nature for me to take over 5,000 pictures throughout the summer. Even if you are not keen on taking photos, I still recommend that you take pictures of everything. I literally mean everything; I took pictures of everything from beautiful architecture and historic landmarks to friends and food. I read my blog and cannot picture my lab, the church down the street or the beautiful lake near my living quarters. Granted, I am not the most descriptive writer, but my journal does not fill me with as many emotions as does each of the pictures of these places.

Each new city I explored, I would make a new computer photo album for. Since most of my travels were day or short weekend trips, I did not want to spend time describing each building or landmark I took a picture of. I found it useful to take a picture of the building's plaque or descriptor, along with writing my route on a map. Since I took so many pictures and was not able to describe or organize them well until months after each excursion, the picture-map documentation helped detail each computer photo album. Taking pictures of food may seem extreme but writing that I ate a bratwurst daily in Berlin does not make it real when I read it today. When I look at my pictures, it is as if I could smell and taste the different foods. I look at the spices and everything on the plate and try to recreate the taste for myself and friends from home. The food picture that caused the greatest stir at home was that of my raw ground meat sandwich. It was unheard of to eat anything close to raw ground meat where I grew up so I took a picture of it as myself and a friend bit into it. It not only captured the existence of the sandwich but our expressions of surprise at it being delectable. Make sure to interact with the environment around you beyond the inanimate objects.

In order to get the most out of your research, you need to interact with other people working in your lab. The more you interact with your lab mates, the more comfortable they will feel with you and you with them. Remember, you are going into their lab and neither you nor they know what boundaries each other have at the beginning of your time together. Be sure to talk with individuals outside of your lab in addition to these inside of your lab. If you try to speak their language, not only will conversations be interesting, your lab mates will most likely respect you more. In general, build as many good relationships with those around you as possible. You may not use the lab techniques again but the relationships and knowledge you gain will stick with you for a lifetime. Also, do not be afraid to ask questions; it shows that you are interested and the majority of people will respect that. Most likely, the researchers in the lab requested to have someone work in their lab (mine handpicked me from dozens of applicants) so they wanted you there. Just remember, do not feel that you are impeding in on their laboratory or research!

The relationships that I valued most were with those people whom we discussed more than the science in our laboratory. We talked about life, culture, entertainment, politics, misconceptions and thoughts about each others' countries. My host lady, for the two-week intensive language program in Berlin, and her American boyfriend drove me around Berlin so I could see the difference between the buildings in East and West Berlin. I heard numerous stories of how life was drastically different on either side of the wall; one being that she was unable get bananas on the East side. My friends in the laboratory also shared stories of the foods they ate and how life changed after the Wall fell. Though they were children, it still affected their daily lives and has an impact on who they are today. We opened up about our childhoods, family get togethers and what we considered our heritage. I knew more about their lives than some of the people that I spent years with in high school.

Their fascination with politics, international affairs and entertainment were the topics of the conversation once those concerning history and family ceased. The one thing that did not change because of the wall was their perception of entertainment. Those in my lab and around town knew more about the American singers and movie stars than I would have known if I studied for weeks. They were up-to-date with all of the facts about the entertainment world and the political world. I found that across the board, they had definitive political views. One of the first questions I was asked by most Germans had something to do with President George W. Bush. The media had a very biased opinion and my friends wanted to know how I actually felt about the decisions George W. Bush had made and if I supported him. There were many misconceptions associated with politics of both our countries. The one cultural difference that surfaced every time we walked through a park or near a beach was public nudity. Though it was legal to lay nude in some parks, my friends did not all agree that it was something that should occur. I had assumed that since they have grown up with it being *just another thing*, that they would support it. Talking about and experiencing their life and culture may have showed my naivety but my knowledge base grew and my relationships with all of these wonderful people had more meaning.

The more freedom you allow for things to occur during your trip, the more you will enjoy your research aboard. Be open to others' ideas and plans and accept things as they happen. If you try to control every moment, you will soon be overwhelmed by the different situations you find yourself in. Remember that your entire trip, not just your time in lab, is designed to be a learning experience. If you take everything you can from each experience you have, your trip will be more memorable. In the end, being open all of the experiences that come your way will allow you to better enjoy your time abroad.

When you have a break in your lab work, go out and explore. You can rest when your abroad session is over. One nice thing about traveling is that the larger area that you explore, the more and diverse experiences you will have. Wander through your immediate surroundings, cities near you and if possible, other countries. Occasionally, go for walks and take in the atmosphere around you. My daily trips to the market and weekly trips into the center of town allowed me to experience the German daily life more. Some of my fondest memories in town were during the Euro Cup, more properly known as the UEFA European Football Championship. For the entirety of the game series, there seemed to be bands and people celebrating Germany's success. The spirit of the town reminded me of Pittsburgh during and after one of our wins during the Superbowl. My travels around town soon expanded into travels throughout Germany and then Europe. I found I enjoyed my travels more when I just went with the flow. The instances in which I planned were only done in order to save money. Fortunately, my program paid me enough to fly there and back plus a little. However, do not let a lack of funding stop you from experiencing the world around you. There are multiple scholarships available that you could apply for to help pay for your time abroad.

Though it is more expensive, I recommend you go out with and travel with the people in your laboratory. You should go to the local restaurants with your lab mates a few times, but keep in mind that you can save money by bringing your own lunch. Strive to immerse yourself in their culture and test the local specialty, whether it be food, art or music. I went to the dormitory that was converted into a massive television studio in order to watch the World Cup. There were fliers up for a Trabant race on a local street. I did not know that it was people pushing these cars from the "olden days" that were made of cardboard, Trabbies. There were local sports days and music festivals that were free to the public. Whether sitting around and just watching all of these things occur around me or actually trying to push a Trabbi, immersing yourself in the happenings around you will make your trip more worthwhile. From the bratwursts in German to the gelato in Italy, every country had their signature staple that should be tried. Things became interesting when I tried the town specialties, such as the raw ground beef sandwiches. There are countless unique things that you will come across if you just take a look around. Traveling can have a steep cost attached to it and some planning in advance will most likely help decrease the expense. Remember that in the long run, an extra $20 here or there is not going to make or break a bank account. If you search hard enough, deals for students and many things you can do for free are readily available. There are free tours in dozens of cities. Remember to take advantage of these opportunities; they will be the experiences that you remember first from your trip. If I could give you no other advice, it would be to take the time to communicate with those at home, in your laboratory and anyone you come across, to make the most of each experience and to take advantage of those opportunities provided to you.

How It Has Impacted My Career

Despite the fact that I am only a graduate student, my research experiences in Germany have already impacted my science career greatly. Not only has it changed many thoughts I had about research, it has enhanced my resume and decide upon graduate school. It has also helped me determine how I would like to run my lab once I become a professor. Immediately upon arrival in Germany, I noticed how accepting people were. I arrived to find my own lab space, equipment and desk ready for me to use. They immediately accepted me as if I were part of their lab and trusted me with equipment. There was no hierarchy within the lab. The Ph.D. candidates were treated just as I, the undergraduate, and the technician were. The hierarchy was apparent when comparing those working in the laboratory and the professor. The professor had worked for many more years to get to her position and was held at a higher standard. Even with this division, we were all close and I was accepted as if I were a childhood friend immediately. On the professor's birthday, she had us all together for a snack to celebrate. We were treated as equals, at least for this moment. Any free time was spent with those in my laboratory, not the other Americans in the program. This was encouraged more in Germany than during my previous summer experience in the United States. My relationships with those in the United States laboratory were more formal. There was less personal information and space shared with those of higher status in the United States. The experience of this different hierarchal environment, immediate trust, and a close relationship with other people working in the lab helped me decide what kind of environment I would like to work in during my graduate studies.

Prior to my research in Germany, I had not seen a lab where a non-formal and seemingly non-hierarchal environment was productive. Previously, there needed to be distinct leadership positions for anything to be accomplished. Our wonderful lab technician did not feel her services were diminished nor did I feel that I was just a summer intern. Everyone accomplished their responsibilities and asked for help without hesitation if needed. Upon experiencing this environment, I decided that once I become a professor, I want to have that type of environment in my laboratory.

Most of my research in Germany consisted of helping my Ph.D. obtain results concerning her Ph.D. dissertation. I learned the techniques to be used in lab by being her "shadow" for approximately ten days. I read a few papers and then was permitted to continue the research alone. I was amazed at how quickly trust was earned in Germany. During my research in the United States, it was not until late-summer that I was able to do everything on my own. The experience level of the student and the complexity of the student's project have a large impact on how quickly a student is able to work independently. I have developed my own sense of how to allow independence of my students in the lab and will be sure to address it on an individual basis.

If someone is not able to communicate their research, it is if it were not even completed. In Germany, we had weekly meetings that consisted of research updates via slideshow presentations. Since the presentations were given in English, I was asked to proofread and critique some of the presentations. The members of the lab developed a deeper appreciation for the English language and I gained an appreciation for how necessary it is to communicate well. This was reiterated when I sat in on a Ph.D. dissertation defense. This was in German but since the slideshow presentation was well organized and used a good variety of graphics, I was able to follow the defense. She was nervous and a little fragmented at first, but in the end, she was able to deliver a great defense. When I defend my own thesis, I will keep in mind that characteristics of a good speaker and a good presentation are universal, no matter what language the presentation is given in.

The main focus of my trip to Germany was to conduct research, but I was never overwhelmed by it. Even when an experiment which took three weeks of preparation completely failed, the upbeat atmosphere of the lab did not change. The lab continued as if nothing failed and we did not allow it to affect their daily lives. More importantly, the members of the lab left their work in lab at the end of the day. This was something that was discussed in the States but I never saw implemented. There were two lessons that I learned from this experience: do not dwell on a failed experiment or let it disappoint you very long and that work, though it can be useful to be discussed outside of the lab, should not take take the place of your personal life. Without a personal life, your work will suffer. Seeing this occur has given me hope that I will be able to react to my research in the same manner through out the rest of my career.

In addition to things that changed my chemistry career, my comfort level in an abnormal environment grew. I had more than just survived my summer abroad, without spending time with any family, friends or a planned itinerary. My confidence and independence soared. Before my summer abroad, I never thought I could be away from my family for more than a few weeks. I had lived no more than an hour and a half from my home my entire life. Being thousands of miles away from everything I knew and was comfortable with, allowed me to considered graduate schools outside a two hour radius. This opened up a myriad of options for me. I applied to four additional Ph.D. programs that were as far as twelve hours away from my home town.

Studying abroad also gave me the confidence to possibly perform my postdoctoral research outside the United States. In Germany, I was able to see various opportunities available to do postdoctoral research in an educational and non-educational environment. I have also continued my search in graduate school for fellowships and international programs. It seems as though the underlying goal of everything I do involves being able to do research abroad again.

The research I did abroad also impacted my current research interests. It helped me funnel them to a specific discipline, biological chemistry. and, at the same time it opened my interests to any research that could be done worldwide. Due to my research in Germany, I now also think that I want to dedicate my chemistry career to research something that is medically based. This would allow me to possibly collaborate with professors in departments outside of chemistry and perhaps overseas.

My experience has opened my eyes to the international community. Overall, research is truly an international effort and should not be separated by language barriers and/or country boarders. It seemed as if everyone in the laboratory meetings spoke a different language outside of their presentations; yet they all came together weekly and communicated. I fell in love with this diverse atmosphere and started to look more closely at graduate schools that had an international community. The opportunity to go to conferences abroad and to collaborate with groups outside the United States became part of my consideration when choosing a graduate school.

For years, I wanted to become a professor, with the notion of teaching at a small liberal arts college. My experience, in Germany, gave me an extra push to become a professor so I would have a larger influence on young adults in the science field. I realized that as a student or postdoctoral fellow, you are only able to influence a smaller amount of people than you would if you were a professor. I want to encourage as many students as possible to study abroad and am now more comfortable with the idea of one day being a professor at a larger school with a diverse community. In the end, my time in Germany helped send me down a path to becoming a professor, which, assuming all goes well, will perhaps become the most important career choice of all.

Chapter 7

How Do You Say "Sulfuric Acid" in Dari?

Patricia A. Dooley*

Division of Natural Science, Mathematics, and Computing, Bard College at Simon's Rock, 84 Alford Road, Great Barrington, MA 01230
***E-mail: pdooley@simons-rock.edu. Phone: (413) 644-4966. Fax: (413) 644-4280.**

In the summer of 2006, Colonel Patricia A. Dooley, Deputy Head of the Department of Chemistry and Life Science, United States Military Academy, West Point, NY, deployed to Kabul, Afghanistan in support of Operation Enduring Freedom to augment the National Military Academy of Afghanistan (NMAA) Implementation Team, a cohort of eight military personnel advising and assisting the nascent academy patterned after West Point. Her mission was to serve as an advisor and mentor to the Head of the Basic Science Department in order to establish a laboratory program and chemistry demonstrations to accompany the one-semester general chemistry course taught to first-year cadets at NMAA. In a country reviving itself after 27 years of occupation, civil war, and governance by the Taliban, and still combating an insurgency, the lack of textbooks, chemicals, reliable utilities, analytical instruments, functioning laboratory space, and instructor experience with experimental procedures demanded a creative response to these challenges.

How Do You Say "Sulfuric Acid" in Dari?

I am interviewing a candidate for the position of chemistry instructor at the National Military Academy of Afghanistan along with the chairman of the basic science division. Colonel Sayeed Akram Raghi asks the candidate some questions, and then it's my turn. I tell the candidate through my translator that I am going to give him the name of a compound and I want him to write the formula on the paper in front of him. Heads nod, task understood. "Sulfuric acid," I say, and my translator turns to the candidate and says, "Soolf-your-ik ah-seed." The candidate correctly writes H_2SO_4 on the paper. *"I got this,"* I tell my translator, and I say in a slow, clear, Voice-of-America tone, "magnesium carbonate." The candidate writes $MgCO_3$. "Sodium phosphate." Na_3PO_4. "Potassium chloride." KCl. My fluency in Dari was limited to "Good morning," "How are you?", "I am fine, thank you," and the candidate's command of English was equally scant, but we both spoke the same second language, chemistry.

In 2003 the Afghan Ministry of Defense and the United States Military Academy (USMA) developed a plan for the creation of the National Military Academy of Afghanistan (NMAA) to produce leaders for the Afghan National Army (ANA). Patterned after USMA, NMAA enrolled its first class of 120 young men in March 2005 (*1*), of whom 84 graduated in January 2009 with degrees in sciences and legal studies, civil engineering, computer science, and general engineering along with commissions as second lieutenants in the ANA (*2*). The second class of 212 cadets graduated in March 2010 (*3*); days later, 600 more incoming cadets began their education (*4*). USMA faculty have served as mentors for NMAA since the plan was developed in 2003, and I had the opportunity to deploy to Kabul in the summer of 2006 to establish a laboratory program and classroom demonstration program for the core curriculum one-semester general chemistry course. This is the story of how chemistry brought educators together from institutions both civilian and military, high school through college, and internationally- and locally-supported organizations.

Opening Doors with Periodic Tables

After the requisite pre-deployment processing and training at the Soldier Readiness Center, Fort Benning, Georgia (Figure 1), some very long flights via Germany, Turkey, and Kyrgystan, and a harrowing ride in a cast-off Mercedes Benz jingle (*5*) bus from Bagram Air Base to Kabul, we arrived at Camp Eggers (Figure 2). Two officers from USMA—a mathematics mentor and a geography mentor—accompanied me, and we immediately linked up with the Military Academy Implementation Team (MAIT), a section of the Combined Security Transition Command-Afghanistan. Led by a U.S. Army colonel, the MAIT was responsible for developing the infrastructure, physical and military training programs, and academic curriculum while serving as mentors to the Afghan Army cadre and instructors who implemented the training and education.

Forearmed with the knowledge that I would be going to Afghanistan, I had two wall-chart size periodic tables and some innocuous laboratory items sent ahead —lab goggles, aprons, gloves, spatulas, stirring rods, a few pieces of glassware and hardware for demonstrations, and things a chemistry professor carries in the pockets of a lab coat—and the boxes of supplies awaited me in the MAIT office at Camp Eggers. Armed with these gifts of good will from the USMA Department of Chemistry and Life Science, I was ready to meet the NMAA faculty who were to be my colleagues for the next two months.

Colonel Sayeed Akram Raghi was the chairman of the basic science division and a chemistry instructor. He directed a faculty of two physics, two geography, and three chemistry instructors: Col. Abdul Sattar, Lieutenant Colonel Janat Gul,

Figure 1. Harmony Church, site of the Soldier Readiness Center, Fort Benning, Georgia

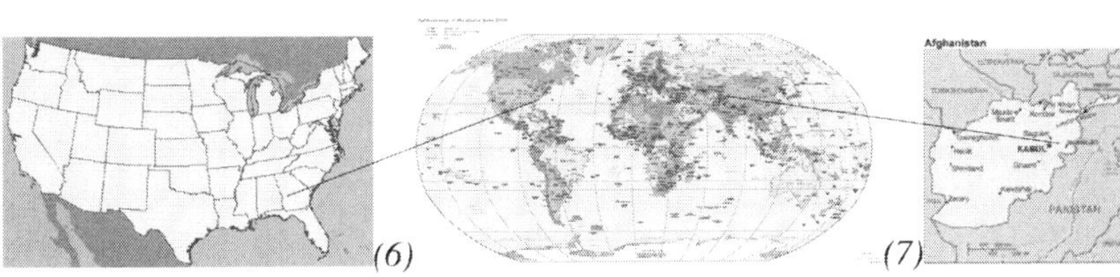

Figure 2. World locator maps of Fort Benning, Georgia (left) and Kabul, Afghanistan (right) (6–8)

and Lt. Col. Jan Daad. Lt. Col. Janat Gul had a master's degree in chemistry, and all four had bachelor's degrees in chemistry. Their taciturn nature made it difficult to determine what they had done since graduating from Kabul University before the Soviet invasion in 1979, but I did discover Lt. Col. Jan Daad had taught chemistry at a Kabul military high school before being appointed at NMAA. Our first meeting in Col. Raghi's tiny office, once a single room in a barracks of the former Soviet Air Academy, was an excited exchange of them thanking me for coming to help, of me expressing my honor at the chance to serve, and all of us attempting to lay out a schedule of what to do next. Their ear-to-ear grins at seeing the wall-chart periodic table were greatly rewarding. After proudly hanging it on the wall of his office, Col. Raghi began pointing to Period 4 elements and reciting their names in our common language of chemistry: "Skahn-dee-yoom, tee-tah-nee-yoom, vah-nah-dee-yoom. . ." This from a man who never uttered a word of English to me (Figure 3).

Textbooks as Manna from Heaven

Anxious to see what the NMAA chemistry storeroom held, I readily accepted Col. Raghi's invitation to examine the chemistry laboratory and stockroom. As the first chemistry mentor, but the sixth mentor to the basic science faculty (physics and geography professors had already trod here), I was pleased to find a moderately equipped chemistry laboratory storeroom Figures 4, 5**)**. Sufficient glassware and equipment were on-hand to run one section of sixteen students through the lab at a time, but the inventory of chemicals was only marginally adequate for instructors to perform demonstrations and wholly inadequate for cadets to perform chemical-consuming experiments. No analytical equipment beyond a pH meter and a balance that measured to 0.1 gram was available. All of the chemicals were donated by the Kabul Science and Technology Center, a state-run organization that trains teachers in the provinces how to teach chemistry and other science subjects; their source of supply was donations from various non-governmental organizations, as I discovered on a visit to their organization later on.

Observing the chemistry faculty teach classes was next on my agenda. I watched Lt. Col. Jan Daad give a class without a translator—big mistake!—and though I could follow the chemical reactions he wrote on the board, without context it was not a very useful classroom visit. I had arrived toward the end of the semester just as the general chemistry course was covering organic chemistry, and the topic of Lt. Col. Janat Gul's lecture was electrophilic aromatic substitution. Sitting in the back of the class with my interpreter whispering in my ear, I could follow the figures he drew on the board, a substituted benzene ring reacting with sulfur trioxide in sulfuric acid to produce the right ortho-, meta-, or para-directed disubstituted benzene. Chemistry was our second language in common.

Watching the cadets in both classes, I was curious to examine what text they were using. I gestured to a cadet to let me look at his book, remembering to open it from the back since the language is read right-to-left. What I saw was a photocopied set of hand-written notes in Dari with the occasional equation or reaction, stapled together to serve as a workbook. No figures, no photos, no graphs, no tables, no images, no glossary, no index, no end-of-chapter problems, no example problems solved in the chapter. The instructors were using the same workbook. Col. Sayeed Raghi had written these notes from his memory of classes at Kabul University, and the MAIT advisors copied and stapled them for distribution to cadets. While this was better than the instructor reading from the only book in the class while students furiously recorded his every word, it squarely contravened the notion that NMAA was patterned after USMA. As their faculty mentor, I knew I had to overcome the shortcoming of no textbooks.

Thinking of my own bookshelf of obsolete editions of chemistry textbooks back at West Point, I mulled over how to empty the shelves of my own colleagues' passé texts from 7,000 miles away. Then the idea struck me: What edition of Zumdahl's *Chemistry* was the current one? Hadn't Professor Zumdahl served on the faculty at the University of Illinois? As Paul Kelter, my ICUC (*9*) compatriot is? Could Paul rustle up some previous editions of Zumdahl's *Chemistry*?

Figure 3. Colonel Sayeed Akram Raghi, Chairman of the Basic Science Division

Figure 4. Inspecting the chemistry and physics stockroom

Dr. Paul Kelter answered my call, collected and shipped twenty used copies of Zumdahl's *Chemistry 6th Edition*, and went back to his colleagues for more textbooks of many subjects. The ICUC was posting my email messages to Dr. Kelter in blog form, and I wrote on July 11, 2006,

"Must report with great exultation that the first two boxes of books arrived yesterday! . . . You should have seen the look of delight on the chemistry instructors' faces when I handed them the textbooks. I got into a shouting match with the basic science department head, however, who wanted to lock the remaining textbooks up in the lab stockroom and forbid the instructors from taking the textbooks home. I raised my voice at him to tell him that if he locks up this knowledge, which must be free and available to everyone, he's as bad as the Taliban. That got his attention!! I then told the other instructors that I wanted to see the Zumdahl book in their possession any time I saw them--at lunch, riding the bus home, after noontime prayers. One very good instructor took my comments to heart—[Lt. Col. Jan Daad] has very little English, but he's teaching himself from Zumdahl now; it's nice that chemistry [is] a universal language (*10*)."

Organic Chemistry Laboratory Experiment

Not only was access to a full-color chemistry textbook with diagrams, charts, photographs, tables, and problems a welcome improvement, the chemistry faculty was very interested in utilizing the chemicals and equipment in the stockroom; however, their most recent exposure to laboratory experiments was a distant memory. Lt. Col. Janat Gul pressed me to devise and implement an experiment pertaining to organic chemistry that all the chemistry sections could do in the regular 55-minute class period. The laboratory room was just a normal classroom with a standing water tank, a couple of small white boards, tall tables used as benches, and stools. No running water, no sink, and only the electrical outlets powered by the haphazard on-site generator. However, the stockroom was equipped with stirring hotplates, magnetic stir bars, burets, and 50-mL Erlenmeyer flasks, and a small amount of ethanol, sulfuric acid, and acetic acid. Olfactory-detected esterification! I researched and prepared a written procedure for the instructors to synthesize ethyl acetate using olfactory sensing as evidence of experimental success. Cadets would dispense acetic acid and ethanol into Erlenmeyer flasks from a buret, heat with a catalytic amount of sulfuric acid for 20-30 minutes, and determine with their sense of smell if the sharp acrid odor of acetic acid had been replaced by the fruity aroma of ethyl acetate, the desired product (Figure 6).

All eight sections of general chemistry taught by three different instructors and supported by this chemistry professor from West Point successfully completed the experiment with no untoward accidental acid spills or hot-plate burns. The acetic acid, ethanol, and the stock of disposable rubber gloves were almost completely consumed. I know I was the highest-ranking lab tech in the Army that day. Getting to chat with the Afghan cadets, whose English was far better than my Dari, was an opportunity I wouldn't miss. The insight I obtained into the behind-the-scenes life of the cadets was immeasurable. They recognized the odor of ethyl acetate as "my sister's fingernail polish remover." A sister who wears nail polish and uses nail polish remover; whose brother, the cadet, knows of these things. . . .

Chemistry at Kabul University: Frozen in Time

After nearly depleting the inventory of acetic acid and ethanol from the NMAA stockroom, I needed to find a source of supply for of chemicals and equipment. Perhaps the Department of Chemistry at Kabul University could give us some, or at least tell us where they ordered their stocks from. A trip to Kabul University was in order, not only for chemicals but for some strategic sowing of good will.

Figure 5. The inventory of chemicals

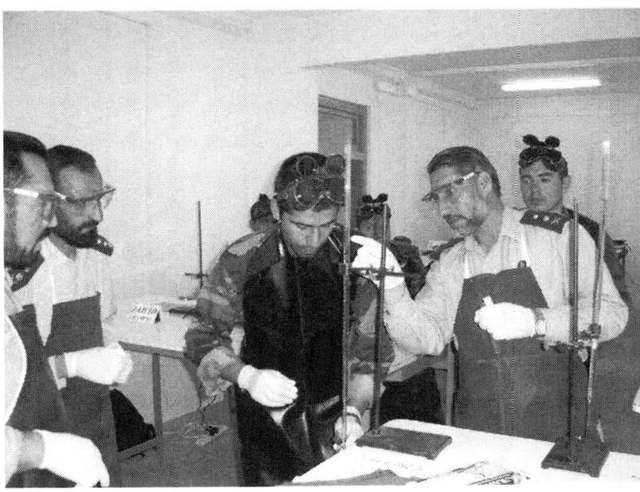

Figure 6. (L-R) Lt. Col. Janat Gul, Col. Abdul Sattar, a NMAA cadet, Lt. Col. Jan Daad in the first chemistry experiment conducted at NMAA

The Ministry of Defense and NMAA were pursuing accreditation as a college degree-granting institution from the Ministry of Higher Education, even though NMAA was in only its third semester of existence. Support from Kabul University was very desirable, and trying to convince the leadership of Kabul University that NMAA was worthy of accreditation led us, the USMA mentors, to visit them and discuss what innovations and progress NMAA was making (*11*). Our first attempt to convoy across Kabul was cut short by the threat of a riot near the university, so we made our way back without incident or success. Our second attempt netted office calls with the dean of the Faculty of Science and Professor

Amiri, the chairman of the department of chemistry, who escorted us on a tour through his department.

My experience at Kabul University further illustrated the common language of chemistry. In the 400-seat lecture hall, the gigantic periodic table painted on the wall above the chalkboard at the front of the room looked like any other periodic table except the last element was number 104 and its symbol, Ku (kurchatovium, instead of Rf, rutherfordium) reflected the influence of Soviet occupation on Kabul University (Figure 7).

A chalkboard in the chemistry laboratory, unceremoniously propped up on a bare tile bench, showed lead(II) nitrate and lead(II) hydroxide reaction equations complete with a down-pointing arrow indicating the formation of a precipitate (Figure 8). While not in written or spoken language, there is universality in a flooded chemistry laboratory floor—especially in a building with no running water, lights, or electricity (Figure 9)!

Kabul University had no chemicals to spare. Their laboratories had been plundered of everything: windowpanes, light fixtures, shelving, drawers, plumbing, electrical outlets (Figure 10). Seeing the great losses this institution had endured made the conditions at NMAA look luxurious in comparison. While we came away with no resources or leads, we extended the good wishes of the American people and the staff and faculty of USMA to them. We provided American textbooks to Kabul University with the promise of more to come; Dr. Paul Kelter kindly arranged for more books to be sent.

International and National Support of High School Education

The search was still on for a source of chemicals and equipment. What other institution in Kabul was teaching chemistry and consuming materials? The answer was right under our noses. Immediately adjacent to Camp Eggers in Kabul is the German Amani High School; a razor wire-topped wall separates the two, and on occasion the security force would open a gate in the wall to let American and international forces run on the school track instead of the treadmill at the Camp Eggers gym. A visit to the school showed the advantages of having generous benefactors (Figure 11) (*12*). We saw students competing in a 'challenge bowl' contest between two physics classes; a lab chalkboard depicting a circuit with three resistors in parallel and the equations to calculate current through each resistor and total current; and a biology class studying human physiology complete with a skeleton and human anatomy model. The subject of the day was the 'degestive system' [sic]. To my dismay, I learned that their sources of chemical supplies were vendors in Iran. I knew better than to attempt a local purchase of materials through Iranian suppliers, even though NMAA desperately needed to obtain chemicals for its own laboratory program.

Since NMAA's only benefactor for chemicals had been the Kabul Science and Technology Center, it was time to visit them and see how flush they were with supplies. Supported by the Ministry of Education, the center educates teachers in the teaching of science, both on-site and in outlying sites. Wending our way through room after room of charts of the laws of physics, diagrams of the

Figure 7. Kabul University chemistry lecture hall

Figure 8. Kabul University chemistry laboratory

electromagnetic spectrum—all in English—and equipment to measure the force constant of a spring or the speed at which a ball drops, the center was a veritable toy store for a science educator. Their chemical stocks were painfully low, and their donors, non-governmental organizations around Kabul, had nothing to give the center. I was dismayed that the only thing I returned to NMAA with was two circular discs of filter paper. But on the bright side, I saw an opportunity for a chromatography demonstration!

Figure 9. Kabul University chemistry laboratory flood

How Do You Say “Sulfuric Acid” in Dari?

The U.S. Army prides itself on developing its leaders to be mentally agile and respond effectively to a variety of situations, and I was called upon to exploit that agility in unanticipated ways. I found myself defending the NMAA curriculum in meetings at the Ministry of Defense, installing a smart board in the chemistry lab, giving faculty development workshop presentations, and interviewing faculty applicants.

In its third semester, NMAA was preparing for an expansion of its corps of cadets from 100 entrants in 2005 to 200 in 2006 and 300 in 2007. The pressure was on to increase the size of the faculty and select instructors to teach courses not yet offered because the first class was only in its second year. One morning the dean’s assistant and a translator stepped into the offices of the MAIT, telling me Col. Hamadullah wanted to see me; in his office, he asked if I would interview a candidate for a civil engineering instructor position and provide him a recommendation to hire the candidate or not.

My first experience with interviewing through a translator was simultaneously frustrating and amusing. Imagine giving yourself twice as long to think about the next question to ask, because the last one is being translated, answered, and then the answer is translated back into your language. I grew accustomed to this lag time, but for the first interview I did I grew impatient waiting for an answer. Interested in the chemistry of concrete, I asked what proportions would he use to mix concrete to build a road or exterior wall or interior floor, and why? As far as I could tell from his answers, there was only one way to mix concrete and that was the way the state of Afghanistan directed! I grew more comfortable with interviewing candidates, every time being summoned by the dean with no prior notice. The foreign language candidates for English instructor were the easiest to interview, the religious and cultural studies candidate the most interesting. Watching one candidate, a graduate of the prestigious Al-Azhar University in Cairo, verbally

Figure 10. Kabul University chemistry teaching laboratory

Figure 11. Amani German High School biology laboratory

spar with the department chairman over an interpretation of the Quran gave me great respect for their intellectual prowess and the depth of their faith in Islam.

The candidates for teaching positions in the department of basic science were the ones I could discriminatingly interview. The vignette at the start of this chapter describes the first interview I conducted with a chemistry candidate, and I drilled all of them with a set of questions that revealed if they knew enough to be able to teach cadets. I found the physics candidates to be similarly fluent in the common language of physics; I would ask them to write the equation for force. "F = m x a" goes on the paper. Work? "W = f x d". Power? "P = W/t". I might not have been able to ask where they grew up or what they liked to eat for dinner, but we could communicate in science.

One particular interview with a chemistry instructor candidate gave my translator and me a good laugh, although at the expense of the candidate. One question I asked separated the more experienced candidates, many of whom had taught chemistry before, from the less experienced ones. Col. Raghi would ask the candidate to identify and describe the three states of matter. "Solid, liquid, gas." No wrong answers there. He would then ask how you would convert a solid to a liquid; a liquid to a gas? "Apply heat." Everyone gets the answer. I would follow by asking how would you reverse the process? "Remove heat." The stumper: How would you convert a gas to a liquid at constant temperature? "Increase the pressure." Those who immediately answered the question were the old pros, but the ones who had to mull it over demonstrated either keen thinking skills to reach the correct answer or they made up new science on the spot. A couple of times my translator had to repeat the question; once I had to write out the equation for $H_2O(g) \rightarrow H_2O(l)$ to clarify the question. One young recent graduate thought and thought, then announced, "It cannot be done." After the candidate left, I turned to my translator and we looked at each other in amazement, broke out laughing, and I commented, "It cannot be done? What, the laws of physics don't apply in Afghanistan?" He replied, "According to him, no, they don't work around here. Maybe you have to go to Pakistan for the laws to apply!"

Calculators Make Training the Trainers Easier

To my dismay, very few of the candidates for the chemistry positions had any laboratory experience, and asking them to explain how to prepare a solution of 1 molar sodium hydroxide was met with silence or blank stares. Was the shortage of chemicals in their undergraduate program to blame? Was their chemistry education all theoretical, not laboratory-based? How was I to knock off the rust from my chemistry colleagues' long-ago laboratory experience?

My mandate was to inaugurate a laboratory program to accompany the one-semester general chemistry course and establish a classroom demonstration program to bring a hands-on, right-before-your-eyes component for cadets to see the precipitation of an insoluble sulfate, the appearance of pink in an acid-base titration with phenolphthalein indicator, or the deposition of copper metal on an electrode in a voltaic cell. In our American college and university teaching labs, we are accustomed to reaching for the 0.10 molar solution of copper(II) sulfate off the shelf to do the dilution demonstration, or for equimolar concentrations of sodium hydroxide and hydrochloric acid for coffee-cup calorimetry experiments. Reaching for the stock solution of any chemical to put on a laboratory experiment or classroom demonstration would be a handy situation for the NMAA chemistry instructors, so I decided to show them how to prepare a solution of a desired concentration. At one of our afternoon lab sessions, I informed them that we would be mixing up a solution of copper(II) sulfate and then diluting it to a desired concentration (Figure 12).

At West Point, this particular exercise was an annual classroom demonstration; cadets calculated how much copper(II) sulfate pentahydrate was required to prepare 250 milliliters of 0.100 molar copper(II) sulfate in a volumetric flask, then determine the concentration after 25.00 mL are removed and placed in a 100.0-mL volumetric flask and diluted to the mark. Selected cadets then perform the measurements, prepare the solution, operate the pipette, dilute the solution, and examine the paler blue color. The demonstration is rich with learning objectives: solution stoichiometry, the dilution equation, and proper techniques for a volumetric pipette, pipette bulb, and volumetric flask. At NMAA, I found the opposite: the demonstration was rife with resistance, reluctance, and argument from the chemistry faculty.

First came the calculation of the molar mass of $CuSO_4 \cdot 5H_2O$ by long hand. The chemistry instructors reached an agreed-upon value. The next step, determining how many moles of copper(II) sulfate were in 250.00 mL of a sample of 0.100 molar concentration copper(II) sulfate, was unbelievably challenging for me to explain. By the time I had shown them how to calculate the mass of copper(II) sulfate pentahydrate we would need, I was mentally spent. When Col. Abdul Sattar asked me at the end of these calculations why we were doing this and what purpose would they ever have to mix a solution, I grasped the chasm of our differences and their resistance to pencil-on-paper calculations of multiplying and dividing five-digit numbers by other four-digit numbers. At that moment I realized these instructors probably had neither done a laboratory experiment for the last 25 years nor ever had the luxury of reaching for the 0.5 molar sodium chloride solution off the shelf, and they would likely not want to if it meant the distraction of multiple two- or three-minute long hand calculations to do so.

Bridging the gulf between what I expected the chemistry instructors to know how to do and their ability to do it was my next challenge. First, they needed calculators to facilitate rapid calculation of otherwise onerous multiplication and division. One of my colleagues, the mathematics mentor from USMA, had remarked that the NMAA logistics officer had a stock of calculators in a supply container. I mentally thanked the unknown mentor predecessor who ordered enough inexpensive basic function calculators for the entire faculty and corps of cadets and cursed the unknown mentor predecessor who introduced the NMAA staff to military bureaucracy. To get the NMAA logistics officer to issue eight calculators to the basic science division faculty I drafted a memo, located an interpreter to translate and type the memo, asked the dean to sign the memo directing the logistics officer to release the calculators, and had each instructor report to the logistics officer to sign the memo saying he had received a calculator and was now accountable for it. Calculating 250.00 mL x 0.1 mole/1000 mL x 249.68 grams/mole was now a lot easier.

To get the chemistry faculty accustomed to the idea of routinely preparing stock solutions as needed, I had to convince them why they needed to have solutions at hand. The work I was doing to develop the chemistry curriculum provided the answer. In response to a tasking from the dean, I prepared a draft syllabus of lesson titles for their one-semester 50-lesson chemistry course that covered the topics of the two-semester course taught at West Point. The chemistry instructors then asked for a list of lesson objectives to accompany each lesson, which eventually resulted in a robust syllabus that included a 3-4 minute demonstration or reference to an animation or video for each lesson. I also prepared detailed procedures for the conduct of eight experiments scheduled throughout the 50-lesson syllabus. Hands-on demonstrations and laboratory experiments would require the instructors to prepare solutions.

Computers Changed the Assessment Pedagogy

Issuing calculators to the science faculty was the first step in demonstrating the computational power of information technology. When the staff realized how much easier it was to compile and manage grades in a computer spreadsheet, the assessment pedagogy changed dramatically.

The commander and dean at NMAA were eager for recognition from the Ministry of Higher Education (MoHE) but had to balance the desire to emulate the curriculum and pedagogy of USMA with the need to avoid implementing an education system so different that it might lose the support of Kabul University and alienate the MoHE. One aspect of education at NMAA that could be improved was the frequency of cadet evaluations. During the first two semesters, the cadets' only graded event was a final examination in every course; no homework, periodic quizzes, tests, or a mid-term were given. During the semester of our deployment, the commander had agreed to permit the administration of a mid-term examination so cadets could determine how well or poorly they were doing. I still flinched at the idea of only two graded events for an entire semester; where was homework in the equation? Short quizzes? Chapter tests? Then it dawned on us that the dearth of grades was directly related to the amount of effort required to manually record and compile grades, then calculate cadet averages and standings. The staff was calculating the grades by hand!

Major Jong Chung, my mathematics colleague accompanying me from West Point, wrote in his trip report, "One day, after my afternoon statistics seminar, the computer science laboratory technician asked me how he could write a simple conditional statement in Excel™, and I showed him how to do it. During my computer class next morning, I gave the mathematics faculty an instruction on how to assign grades for the test scores using a simple conditional statement in Excel™, and they found it fascinating. The word reached the Dean through Col. Dooley, and the Dean requested that I give his staff a class on how they could manage the cadet grades using Excel™, which would be an essential part of maintaining cadet academic transcripts (*13*)."

To convince the commander and dean that more frequent student assessment was the norm, not the exception in the United States, I hastily conducted a web search of national and local colleges for a sampling of grading policies in engineering, science, and math courses. Course web sites from the University of Texas, Boston College, Cabrillo College in Aptos, CA, and Princeton University showed a range of 15% to 50% of various course grades based on final exams and the remainder on homework assignments, projects, papers, and exams. Empowered by the ability to leverage a computer program to maintain grades and the knowledge that "the way they have always done it"—a final exam grade only—was no longer necessary, the commander and dean had the flexibility to administer as many graded events as they pleased. In this case, mathematics was the common language, and showing the faculty how to speak it more easily with the help of a computer program made the chore much easier.

In the end, the chemistry laboratory enjoyed upgrades of a periodic table and a smart board to replace two small white boards (Figure 13), the instructors received calculators, chemistry textbooks, computer disks full of lesson plans, demonstrations, laboratory experiment procedures, an invitation to join the ICUC, and the undying respect of this professor. All we have in common is the language of chemistry and the desire to effectively teach chemistry to the future leaders of our nations. Maybe that is enough.

Where Are They Now?

The alumni association of NMAA consists of fewer than 300 graduates from the classes of 2009 and 2010 as of this writing; they serve as commissioned officers in the Afghan National Army. The current four classes are populated by 1,300 cadets, 70 of whom have been selected to attend medical school after graduating. The first cohort of women entered with the class of 2013. The 640 cadets of the Class of 2014 were selected from more than 3,000 applicants, the largest group to take the entrance examination in November, 2009. Mentors from the United States Military Academy, the United States Air Force Academy, the Turkish Army, and the Indian Army continue to support NMAA.

The outlook for the Basic Science Division is slowly improving. While NMAA has not yet found its own source of supply for chemicals, U.S. mentors recently ordered $8,000 worth of chemicals and $67,000 worth of chemistry supplies and equipment, both of which arrived in March 2010 (*14*). Having these supplies will enable more than rudimentary chemistry laboratory experiments to be performed. Without textbooks in Dari, however, implementing the syllabus I prepared is challenging, but the instructors are making an effort to do so. Translating a mainstream general chemistry textbook into Dari remains to be done, at which point the rigor of chemistry classes could be improved. The commitment of the U.S. mentors to the success of NMAA is solid; there will come a time that the Afghan faculty finds the motivation to develop their own initiatives and fulfill the NMAA's reputation as 'the crown jewel of the Afghan National Army' (*2*).

I retired from the Army in 2008 after a 30-year career as a Signal Corps officer and chemistry professor, at which time I took a position at Bard College at Simon's Rock, Great Barrington, Massachusetts, where I teach chemistry at the nation's first early college. Small classes of less than twenty highly motivated, eager learners with intense curiosity about science and the world around them are characteristic among NMAA, USMA, and Simon's Rock. Hunger for knowledge and understanding is common to the cadet and student; the Afghan cadet who hung on my every word I used to explain electronegativity was as enthusiastic as the USMA cadet learning stereoselectivity in organic chemistry addition reactions and the Simon's Rock student who wanted to know how those extra neutrons got into the nucleus of an isotope. I employ the Socratic method with Simon's Rock students, just as I did with cadets, rather than lecture; when students have no

Figure 12. Training session at NMAA chemistry laboratory

Figure 13. Lt. Col. Janat Gul in front of the chemistry laboratory room periodic table and smart board

more questions for me in class, I start asking them questions about the assignment to probe their level of comprehension. My experiences with military cadets prepared me well for engaging students of diverse backgrounds with a wide spectrum of preparation for college chemistry—the early college student leaves high school after 10th or 11th grade. More importantly, I realize that my respect for the perspective, circumstance, and goals of each student or cadet is vital to his or her success. Most importantly, I strive to impart in the student or cadet the ability to perform critical analysis and rational thought for decision-making and problem-solving, a crucial outcome of higher education in any venue, military or civilian. I am honored to do so and to have done so in both environments.

References

1. Peat, R.; Rice, F. Afghan Military Academy Opens Gates to Future Leaders. http://www.globalsecurity.org/military/library/news/2005/03/mil-050328-afps03.htm (accessed March 13, 2010).
2. Newborn, T. National Military Academy of Afghanistan Host 1st Graduation Ceremony. http://www.cstc-a.com/index.php?option=com_content&view=article&id=173:national-military-academy-of-afghanistan-host-1st-graduation-ceremony&catid=61:january-2009&Itemid=151 (accessed March 13, 2010).
3. Maier, M.; Davis, M. President Karzai Congratulates Future Leaders. http://www.ntm-a.com/news/1-categorynews/330-president-karzai-congratulates-future-leaders?lang (accessed 24 March 2010).
4. Soldiers to Get New Cammo Pattern for Afghanistan. http://www.ledger-enquirer.com/2010/02/25/1030208/02-26-2010-around-the-army.html (accessed March 13, 2010).
5. Jingle trucks are so-called because of the jingling sound of chains hanging from the vehicle; painted with bright, intricate patterns and designs and adorned inside with tassels, they provide colorful relief in a desert country. Buses were similarly adorned.
6. Map Maker. http://www.nationalatlas.gov/natlas/Natlasstart.asp (accessed March 24, 2010).
7. Political Map of the World, June 2009. https://www.cia.gov/library/publications/the-world-factbook/maps/refmap_political_world.html, (accessed March 24, 2010).
8. Afghanistan. http://www.state.gov/cms_images/map_afghanistan.jpg (accessed March 24, 2010).
9. International Center on Undergraduate Chemistry. http://icuc.unk.edu/english/eng_home.htm.
10. Dooley blog. http://icuc.unk.edu/english/dooley_blog.htm.
11. The pursuit of accreditation was premature. NMAA attempted to gain accreditation by an international body, but the requirement was that NMAA must graduate its first class before applying.
12. The German Amani High School in Kabul was founded in 1924 and supported by faculty and staff from Germany through 1985. Looted and destroyed under the Taliban, it closed in 1992, but German support resumed in 2002; the refurbished school boasts state-of-the-art laboratory facilities. The keys to the school were presented to the Afghan minister of education in a ceremony on September 22, 2003. Press Briefing by David Singh, Public Information Officer, Media Relations Officer, Office of Communication and Public Information, UNAMA. http://www.un.org/apps/news/infocusnews.asp?NewsID=609&sID=1 (accessed March 24, 2010).

13. Chung, J. Memorandum, September 4, 2006: End of Tour Report, Mathematics Academic Advisor to National Military Academy of Afghanistan (NMAA); June 2–July 27, 2006.
14. Leon, Robert, Colonel . MAIT Senior Academic Mentor. E-mail message to author, March 15, 2010.

Chapter 8

The ICUC and the Benefits of an International Chemistry Education Organization

Carlos M. Castro-Acuña[1] and Paul Kelter*,[2]

[1]Laboratorio 112 Posgrado, Electroquímica de Semiconductores Facultad de Química, UNAM, Coyoacán, 04510 México D.F.
[2]Department of Teaching and Learning, Northern Illinois University, DeKalb, IL 60115
***Phone: (815) 752-0327. Fax: (815) 753-8594. E-mail: pkelter@niu.edu.**

Individuals with a very different cultural background but with a common enthusiasm for teaching chemistry in the best way possible and the desire to share their points of view and experiences along their academic careers are the main ingredients to create a successful center of information where everyone can benefit from each other member. ICUC is the answer for many academic needs that teachers share worldwide particularly when they are in charge of teaching the first course of chemistry to youngsters that are not necessarily interested in the subject; despite this lack of interest, teachers will try to give them the basic chemistry tools to become citizens that will use this knowledge to take better decision. In this article we show some testimonies by Chemistry teachers from different countries about how a center like ICUC serves as a great option to get better results in their day to day teaching activities.

Personas con una formación cultural muy diferente pero que comparten el entusiasmo por enseñar química de la mejor manera posible y el deseo por compartir sus puntos de vista y las experiencias que han acumulado a lo largo de su carrera docente, son los principales ingredientes para formar un centro de información donde cada miembro pueda beneficiarse de todos los otros miembros. El ICUC es la respuesta para muchas de las necesidades que los profesores de todo el mundo tienen, especialmente cuando tienen a su cargo el primer curso de química con jóvenes que no están necesariamente interesados en esta asignatura; a pesar de esta falta de interés, los profesores tratarán de enseñarles los conceptos básicos de química que les permitan convertirse en ciudadanos que puedan utilizar sus conocimientos de química para tomar mejores decisiones. En este artículo mostramos algunos testimonios de profesores de química de diversos países acerca de cómo un centro como el ICUC les ha servido como una muy buena alternativa para obtener mejores resultados en su labor docente cotidiana.

Introduction: How an International View of Knowing Sets a Basis for Working Together in a Worldwide Organization

Culture (from Merriam-Webster (*1*))
Pronunciation: \ˈkəl-chər\
Etymology: Middle English, cultivated land, cultivation, from Anglo-French, from Latin *cultura*, from *cultus*, past participle.
Definitions

A. The integrated pattern of human knowledge, belief, and behavior that depends upon the capacity for learning and transmitting knowledge to succeeding generations.
B. The customary beliefs, social forms, and material traits of a racial, religious, or social group; *also* : the characteristic features of everyday existence (as diversions or a way of life) shared by people in a place or time <popular *culture*> <southern *culture*>.
C. The set of shared attitudes, values, goals, and practices that characterizes an institution or organization <a corporate *culture* focused on the bottom line>.

Our own sense of "culture" is the shared values, attitudes and behaviors that define a group or, more broadly, a society. Each of us is a product of our own culture or, perhaps more accurately, cultures, because we are defined by so many cultural groups, each of which influence us. The scientist in each of us might express this as:

$$\textbf{Each of us} = \sum [a^{*}\text{culture}_1 + b^{*}\text{culture}_2 + \ldots .i^{*}\text{culture}_n]$$

While we cannot quantify the specific impact of each culture, we can appreciate its influence as a part of who we are.

Among the groups of which one of us (Paul B. Kelter, PBK from now on) is a member, either by birth or by choice, include: U.S.-born; male; White; 55 years old; Brooklyn, New York; Jewish; mom was a bookkeeper; dad fixed air conditioners and TVs; public school-educated; grew up in urban housing projects; went to summer camp; international folk dancer; runner; divorced parents; wife from Nebraska; chemist; chemistry educator. CMC-A's cultural connections include: Born in Costa Rica; male; White; Mexican by heart and by marriage to a beautiful Mexican; 56 years old; Catholic; private school precollege education but public university education; divorced parents; travel fan and dog lover. These cultural influences have a profound cumulative affect on our individual values, those closely-held ideas on which we base our decisions. Taking just three of PBK's list of cultural influences can help explain some of his own values, shown in Figure 1, in which where he lived, his lifelong hobby, and the culture in which his wife was raised are key influences.

As chemists and chemistry educators, we share common cultural influences of both education and science that interact with our individual cultural influences as each of us develops our approach to chemistry education. In 2007, the staff of the journal *Science* interviewed 12 prominent science faculty from 6 continents about their view of science education at the postsecondary level (*2*). With whatever differences they had as individuals, as members of the community of scientists and faculty, they noted themes that are common to their perception of the culture of students, education, and society, including: lagging interest and poor preparation among students; excessive professional burdens; and anti-science attitudes among society at large. *The sense of the prevailing culture that these scientists share, along with their own more regional and personal cultural influences lead to the values that they take into the classroom, and the resultant ways in which they teach their students.*

There are those whose work points to the presence of a national culture of education, as evidenced by national norms of educational processes. A strong argument is made for this in *The Teaching Gap*, the 1999 book by .science educators James Stigler and James Hiebert, who analyzed hundreds of middle school mathematics classes in Japan, Germany and the United States, noting cultural differences in the ways in which teachers are trained and classes are taught among the three countries (*3*).

The bottom line is that as we develop our *philosophy of teaching* – our set of values that guides us both in our work with our students and our professional development – *we can see it, in large part, as a product of our cultural influences, our understanding of our own and our students' place in this culture, and our understanding of the role of education (especially chemistry education) in this culture.* Because these many cultural influences are constantly shifting with time as individuals and societies change, our philosophy of teaching must continue to either shift or be reinforced by these changes. In no case, however, should we permit it to stagnate without periodic attention.

In many parts of the world, the educational practices that express the cultural influences in the United States are often used as the models for education. More bluntly, the U.S. is often seen as the exemplar of educational practice, and several countries reward faculty who publish in U.S.-based journals with significant salary bonuses that, in some cases, can be even higher that their annual salary; this happens, for example, in Mexico with the Universidad Nacional Autónoma de México's (UNAM) bonus program. Research-based faculty can get an additional bonus with the National System of Researchers (SNI) program . Other countries, such as China have sent groups of faculty to the U.S. on a regular basis to learn about alternative models of instructional practice. PBK worked with several such groups for one-semester programs as part of the University of Illinois Intensive

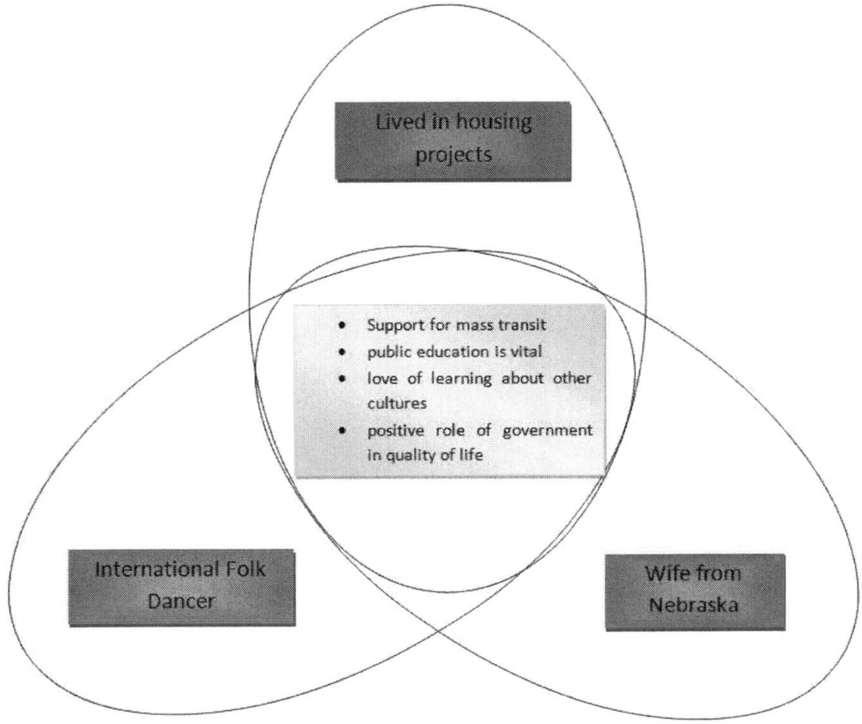

Figure 1. PBK Cultural Influences and Outcomes

English Institute (*4*). CMC-A considers that the time spent at The University of Texas at Austin completing his Masters and Ph.D. theses thanks to a Fulbright Fellowship gave him a tremendous advantage in getting a full time position at UNAM's Chemistry School.

Can we, and, in fact, *should* we, build toward a common, worldwide, philosophy of chemistry education that does not *a priori* place the U.S. at the top of the pyramid, but, rather, recognizes the *common culture* we share as chemists and the common sense of the purpose of chemistry education *across borders*? This chapter is called, **"The ICUC and the Benefits of an International Chemistry Education Organization"**, because the organization that we are about to describe, the International Center for First-Year Undergraduate Chemistry Education (ICUC) (*5*), *is based on the fundamental construct that chemistry teachers from all countries and cultures have made important contributions to our global educational practice, and we must continue to encourage and honor this worldwide work.* The ICUC members recognize the mutual intrinsic intellectual and extrinsic professional benefits that we gain from sharing ideas as a way of enhancing what we know about each other, the way in which we practice chemistry education, and the constantly changing cultural influences on that practice. *The organization specifically rejects a hierarchy in which one country is seen as more educationally savvy than another*, which has so often resulted in a unidirectional flow of ideas ("We teach, you learn"). Rather, it asserts that communication is a multi-way street, in which ideas are treated with dignity, irrespective of the source.

Taking this one step further, in this educational worldview, we actively seek the ideas of those whose culture is different from our own, because it helps to build in us the broadest possible perspective about how to educate. We are each products of our experiences, which are often framed in cultural norms. Enhancing our understanding via working with, and learning about, the experiences of others expands our understanding, and therefore, our professional capability, as chemistry teachers. *This chapter is about the formation, outcomes and impact of the ICUC as a model of international cooperation in chemistry education across cultures.*

History of the ICUC

The authors first met at Chem Ed '91, a conference held at the University of Wisconsin-Oshkosh (UWO) in late May 1991, and organized by PBK, who was, at the time, in charge of the Science Outreach program at UWO (*6*). The conference attracted over 1,000 high school, college and university faculty from the U.S., Canada and Mexico. We each were roughly 10 years into our academic careers, and found that we had common educational concerns and interests. We also became fast friends. During the past 19 years, we have worked together on many professional initiatives, including chemistry education papers (*7–22*), conference presentations in Latin and North America (*23–47*), and some curriculum development. The collaboration was (and is!) pleasant, and educationally and culturally rewarding.

The idea for an international organization devoted to the common interests of teachers of first-year university-level chemistry was conceived by PBK during an afternoon run (such runs are great for letting the mind wander, ending up in the most useful places!) in April 2003, a month before he began a position as Director of General Chemistry at the University of Illinois at Urbana-Champaign (UIUC). This large state university had the resources to allow such big dreams to take root, at least in a limited way.

After some discussion, the authors decided to have a meeting on the UIUC campus, and, because of limited funds - about $20,000 - invite colleagues from the U.S., along with CMC-A from his home in Mexico City. The resulting mini-conference included 25 "faculty" (we use the term for all professionals who teach at the college and university level) who share a passion for chemical education and whose major teaching assignment was the large lecture first-year chemistry course.

This meeting begat the International Center for First-Year Undergraduate Chemistry Education, or ICUC, pronounced "E-cook" in Latin America, where we now have many members, especially in Mexico, Argentina, Chile, Uruguay, Venezuela and Peru. We like to link ICUC with the acronym "I see, you see," because this is the main idea, to share a common vision of our world based on the combination of individual perspectives. E-cook now has about 200 members from countries on 6 continents. Our current Board of Directors includes faculty from Bulgaria, Mexico, Argentina, the U.S. and Peru.

Goals of the ICUC

The ICUC strives to meet as many professional chemistry education needs as possible for its members. To the extent that our time and resources allow, we seek to:

- **Provide Resources.** Share the knowledge and skill of faculty/ professional staff with our colleagues world wide. Serve as a clearinghouse for information pertinent to first-year chemical education.
- **Content.** Consider which concepts and topics are included in first-year courses and what should be included in such courses. What should students learn from the courses?
- **Focus On Learning And Communicating.** Actively participate in the research, development, testing, assessment, and dissemination of information appropriate for first-year chemical education.
- **Sponsor Learning Opportunities.** Host regional, national and international conferences/workshops.
- **Enhance Available Technology.** Organize, catalog, and prepare streaming videos of demonstrations, laboratory experiments and lessons. Provide access to state-of-the-art molecular modeling, computer simulations, animations and drill and practice materials for use in first-year chemistry.
- **Create Forums to Promote Professional Growth.** Actively support faculty in chemical education.
- **Sharing Through Writing.** Provide an international forum for disseminating information of interest to first-year via newsletters and journals.

Accomplishing the Goals: The Challenge of Worldwide Communication

The organization has no budget. It has grown based on the appeal of its goals, the goodwill of its members, and the creative use of available resources at the host institutions, which are, for the primary Website, The University of Nebraska-Kearney and, for the listserv, Northern Illinois University. The life blood of any organization is communication. The ICUC seeks to stay in communication with its international audience via a listserv, conferences, newsletters, and, most recently, on-line forums.

The ICUC had two conferences, the inaugural one in May 2005 at the UIUC campus in Illinois, and a second one at the University of Colorado-Boulder in May 2007. Both three-day conferences attracted a worldwide audience of well over 100 educators, and included symposia on items of mutual interest, including, from the 2007 Boulder meeting:

- Engaging Students in Global Issues;
- Partnerships between Community Colleges and Universities;
- New Ideas and Techniques in Lectures and Labs;

- Chemical Education Technology;
- Environmentally Responsible Chemistry, and;
- Chemical Education Research

Reviews of both conferences were nearly universally positive. Additional conferences have resulted because of the international connections fostered by the ICUC. The 15th International meeting of the Proyecto Integral de Educación Química – the Integrated Chemical Education Project – was held at the Universidad de San Luis in Argentina in October 2006 with members of the ICUC as well as other chemistry educators throughout Argentina. In July 2007, a conference sponsored by Real Sociedad Española de Química attracted over 250 participants, including a dozen ICUC members from as far away as South Africa, was held at Universidad Politécnica de Madrid in Spain. The main promoter of ICUC in Spain is Dr. Gabriel Pinto-Cañón, whose work has been vital to the success of the organization, as we will discuss.

We give certificates of recognition to the top first-year chemistry student in each member's institution, as well as runners-up, as named by the member. Other than postage, the certificates cost us nothing but a little clerical time, and the meaning to the students and the faculty member are great. We used to give small checks, and then textbooks from a sponsoring textbook company, but with the worldwide recession, these extra prizes are no longer possible. We are in the seventh year of the certificate program, which is valued by our international and U.S. membership.

Newsletters that included contributions about chemistry education from members throughout the world were published on-line (*48*) for 3 years. However, with the lack of institutional help at PBK's current professional home, Northern Illinois University, the newsletter has been suspended. In its place, we have Ning-based Communities of Interest, (*49*) in which members share and discuss resources and ideas concerning:

- Applying Real-Life Topics in First-Year Chemistry;
- Laboratories in First-Year Chemistry;
- Problem-Based Learning;
- First-Year Chemistry in Secondary Education;
- Cultural Diversity in First-Year Chemistry, and;
- PEI - Integrated Educational Projects for Science and Technology

The communities are new, and we look forward to assessing their effectiveness.

The ICUC has had some success with its goals of sharing ideas and content, enhancing technology and, especially, creating forums for professional growth. At American Chemical Society conferences and Bicentennial Conferences on Chemical Education, there is often an ICUC-based symposium. One such symposium, conceived of and organized by Charity Flener Lovitt, led her to propose this book, for which she is the primary editor.

Building an organization with nearly no funding requires the drive and commitment of its members. We do not charge dues, because we do not wish to have economics dictate participation. This is particularly important for teachers from many countries where the per capita incomes are quite low, and we know that teachers are often at the bottom of the academic salary ladder. We wish to have a level economic playing field. Unfortunately, this means that we have no predictable funding base. We have been fortunate to grow via people telling people about the ICUC. *This networking has been especially strong in Latin America and Spain, and the growth of the organization's work in these areas is an object lesson about how to build relationships on behalf of science – and all – education.*

The ICUC in Latin America and Spain

When ICUC was created, one of the main aspects was the special interest in including chemists from Spanish-speaking countries, in order to form a strong community among teachers with *a different spoken language, but a common chemistry language*. To describe how the ICUC has been growing in Latin America and Spain, we need to give some background.

We noted earlier that a group of faculty met in October 2003 at the University of Illinois at Urbana-Champaign, agreeing at that meeting to form an international organization that is the ICUC. Among the most difficult discussions was the range of courses that would be part of the entry-level universe.

We, and our colleague, Cathy Middlecamp of the University of Wisconsin-Madison, wrote about that discussion,

> "...identifying a single common name for the group of courses that we all teach was problematic for 25 instructors who had 25 sets of experiences for, in several cases, more than 25 years! World-wide, many names are in use:
>
> - Preparatory chemistry. (For instance, in Mexico, the chemistry courses in high school are often more similar to many first-year chemistry courses in USA than the general chemistry course that we teach at UNAM's Chemistry School.);
> - chemistry for X, where X includes chemistry majors, non-science majors, biology majors or engineers, etc.;
> - general, organic and biochemistry (GOB);
> - honors-level chemistry, accelerated chemistry;
> - environmental chemistry;
> - chemistry in our world.
>
> In terms of what we teach, we recognize the individual flavors among the tens of thousands of postsecondary institutions around the world (including 3500-plus in the U.S. alone.) However, to one degree or another, we agree that each of these courses considers many areas of chemistry so that *a student leaving any of these courses should (arguably!) be able to intelligently discuss certain fundamental ideas of chemistry and how they apply to each of the traditional areas of our*

subject, if not to wider issues of importance to that student and/or to our society" (*20*).

We also noted in that article *the unity of the questions all committed teachers of entry-level chemistry courses face, no matter what country they call home.*

> "As professionals who focus on what we will henceforth call "First-Year Chemistry," we have common interests in the key areas of the curriculum that apply to this field. Here are some of the questions to which we seek answers as we design first-year chemistry curricula: (Note: If we could ask these questions to Latin American teachers we would be surprised with the diversity of responses. Every country has its own educational system and there is practically no contact with the systems of other countries. In this context, an organization like ICUC has many things to offer in order to foster and maintain a communication network.)
>
> 1. How do we balance the multiple purposes of the first-year chemistry course?
> 2. To what extent do we connect to topics and issues beyond our discipline?
> 3. How do we determine which students are prepared for our courses?
> 4. In what ways should we use the classroom time allotted to us and, more broadly, what research supports the choices we make?
> 5. How do we evaluate what students know during the class period, during office hours and outside of class via assignments, examinations, presentations and other tools?
> 6. How might we work with students who are having difficulty in learning?
> 7. What should be our role in promoting success for traditionally under-represented students
> 8. What norms exist for assigning grades to first-year students?
> 9. How can we help each other improve our teaching?"

This article was also published in Spanish to show the strong commitment that ICUC has to inclusion of communities that speak different languages. Since its inception, the ICUC webpage (*4*) has had three versions: English, Spanish and Chinese.

To establish a relationship with colleagues whose language is not English, and whose culture is not U.S.-directed, Mexico was the first natural option due to the physical proximity to the U.S. and the huge penetration that Spanish as a second language has in a vast area of the U.S. that extends well beyond the Southwestern border states, mainly due to the large increase in immigration from Mexico and many other countries over the past few decades. It is noteworthy that several members of the first ICUC Board of Directors were fluent in both languages, and this is still true with our fourth Board of Directors.

The interaction between English and Spanish has not been happening in just one direction. Although the extent of Spanish use in the U.S. is a relatively recent phenomenon, Latin American schools have been using chemistry texts in English for many decades, mainly in the graduate and postgraduate levels, but also at high school and secondary levels, because *for many Latin Americans an education with a strong English component is a key for a better future.*

CMC-A's work as the ICUC's Director of Latin American Affairs, extends to his relationship with colleagues in Europe, especially Spain. Soon after the founding of the ICUC, CMC-A asked Gabriel Pinto Cañón, an internationally known chemistry educator at the Department of Industrial Chemical and Environmental Engineering of the Polytechnical University of Madrid, to join the organization. Dr. Pinto Cañón turned out to be a wonderful recruiter for the organization, widening the scope of the organization via his work with colleagues in South Africa, Italy, Bulgaria and other countries who are active members of the ICUC.

Thoughts about Relationships and the ICUC from Our Iberoamerican Members

The importance of the ICUC as an organization that recognizes common professional goals, and the need for a common system of support among educators worldwide can be highlighted by the thoughts of our members from Latin America. As educators, we say that this *qualitative form of assessment* represents a powerful statement of the impact of the organization. As people who wish to highlight the relationships, we say that these stories talk to the intrinsic good of scientists sharing our humanity.

I. Carlos M. Castro-Acuña

"I was very lucky to be a member of the First Board of Directors and felt even more so when I was appointed the Director of Latin American Affairs. At that time, my main goal was to work hard to have more members from Latin America, and also from Spain, because this country has had a tremendous impact in all aspects of our education.

"Here was my official presentation to the faculty attending the founding meeting of the ICUC, October, 29, 2003. This was the first such contact most of the attending faculty had had with a faculty member from Mexico:

> 'In Latin American countries sometimes we have very long names and use both last names. My full name is Carlos Mauricio de la Cruz Castro Acuña. I have been a full-time teacher since 1980 at Facultad de Química, Universidad Nacional Autónoma de México (UNAM). UNAM has more than 260,000 students on several campuses. In the Chemistry School at the main campus, we have around 4,500 students, all pursuing chemistry-related careers. My Ph.D. is in Physical Chemistry (Electrochemistry) and I have always been very interested in Chemical Education. I am

one of the organizers of the Mexican Chemistry Olympiads and I enjoy producing puzzles that mix Chemistry and logical reasoning.'

"Since 1991, Paul (co-author PBK) and I have had a very productive interaction where we both have learned a lot not only from our personal experiences but also, despite the profound differences between the U.S. and Mexican educational systems, about the many common issues that we share in our everyday duty as chemistry teachers, and the amazing similarities in the obstacles that we face in order to get better results from our students.

"In one presentation we commented that Mexico and the U.S. are not only distant neighbors in political matters but also in educational topics where we waste many opportunities for having productive interactions. Undoubtedly, one of the main barriers is the language, but unfortunately, another big obstacle is that many people from the U.S. are either not interested in visiting a country that they perceive as unsafe, or do not believe that they have something to learn from teachers from an underdeveloped country.

"On the other hand, in Latin America, we typically believe in many myths, among these is that in the U.S., all universities are very good, all teachers are not only experts in their discipline, but also have a strong pedagogical background, and that all secondary schools have very well-equipped laboratories.

"The only way to have a more real vision of our country compared with other countries is to take advantage of organizations like ICUC that specifically promote the interaction among teachers from different backgrounds in open environments where everyone can speak freely about the advantages and disadvantages of their educational systems.

"For instance, at UNAM's Chemistry School, we have 1500 new freshmen every year, and we need to give them a high-quality education with very few resources. We depend almost entirely on government funding because the students do not pay a significant tuition. (It is about 2 U.S. cents!) However, we decided against having large lectures, with groups of more than 300 students gathered in auditoriums, a very common format in U.S. universities. We tried that in 1971 and we decided it was not a good idea.

"For many years, I have been working with Ramiro Domínguez-Danache, a physical chemistry faculty member at UNAM's Chemistry School, and in 1998, we started to work with PBK on issues related to chemistry education, and have been fortunate to have produced many joint publications. This is a clear example of how it is possible to have a very productive relationship when all participants are willing to overcome the natural language barriers and to share in a very open and honest way their knowledge."

To build a multinational working team in the ICUC, Paul and Carlos decided to invite more teachers. The next natural member was José Miguel Abraham, (JMA), a long-time faculty member from Universidad Nacional de San Luis, Argentina. CMC-A's relationship with JMA goes back to 1988 when JMA invited him to a conference: PRAIMEQ I (Primera Reunión Argentina e Internacional de Metodología de la Enseñanza de la Química) in San Luis that had many goals, including the genesis of a journal that could help to communicate to chemistry teachers in Latin America area in our own language: Anuario Latinoamericano de Educación Química (ALDEQ).

II. José Miguel Abraham

"I was born in Buenos Aires, capital city of Argentina, March 21st, 1944. My family is Catholic Lebanese. I got my degree not only as Chemist (1966) but as Chemistry Teacher (1969) and College Teaching specialist (1998) as well. I have always taught that chemistry can contribute to the preservation and/or recovery of the environment in its natural, social and human aspects. This is why I created the 'Integrated Educational Projects' (*50*). Another big project was to create ALDEQ, now entering its 23 years of continuous publication. To be involved with the ICUC since its beginnings is something that has given me a lot of satisfaction and the opportunity to increase my knowledge and to share my ideas with teachers from all around the world. I have the privilege of being one of the Founding Board Members of the ICUC and also to represent Universidad Nacional de San Luis, Argentina at the ICUC."

José Miguel has added an important dimension to chemistry education for more than 30 years; the necessity to give a social, cultural and environmental framework to chemistry in order to teach this science with an eye toward the development of a society in which the students use their acquired knowledge to solve real social problems. Considering the systemic economic, political and cultural crisis in so many Latin American communities, José Miguel considers it unacceptable to teach chemistry separated from its use as a tool to promote human development. This is the origin, in 1992, of the now well-known IEP "Integrated Educational Projects" or PEI (Proyectos Educativos Integrales). At that time, JMA wrote: "PEI is a new methodological-didactic alternative in science education, especially chemistry, with the main goal to reach an equilibrium between Nature, Humanity and Scientific/Technological development that is especially useful for underdeveloped countries".

Years later, after the formation of the ICUC, JMA saw the intimate connection between his work and that of this nascent organization. "The ICUC is a hope to build a concrete, open and serious framework to assure the participation of everyone who has the conviction that chemistry is a "core science" that can promote natural, social and human development and (to help) regain human values, especially solidarity".

JMA's collaborative work continued with the ICUC. At the 18 Bicentennial Conference on Chemical Education in 2004, we had one of our first ICUC-based presentations that we also consider the beginning of what we call "The Multi-National Project: Chemical education: common Philosophy and Programs along USA, Mexico and Argentina" Within this project, we have produced outcomes that have led to several articles and presentations in different meetings.

José Miguel Abraham has been a leader in the ICUC, earning its 2007 "Frank Torre ICUC Award for Distinguished Service to the Chemical Education Community." He has also been a major recruiter for the organization, continuing to enhance the cultural diversity of the group via enlisting nearly two dozen new members from Argentina and Uruguay.

III. Adela Castillejos Salazar

"The International Center for First-Year Undergraduate Center has been for me a wonderful place where I have found a lot of chemistry teachers who provide me with information and expertise to improve my practice as a general chemistry teacher.

"I had the opportunity to participate in the first ICUC- FYI-Chem Conference (2005), and that was my first experience with the people of the organization. I really enjoyed that conference, the organization, the people that participated, the great University of Urbana-Champaign, and the lot of concepts and experiences that I had with Chemistry teachers from different places of the world. The hosts surprised me with great qualities, both personal and professional. During the 2nd ICUC FYI-Chem Conference (2007) I had similar great experiences in Boulder, Colorado. There, Margaret Asirvathan (University of Colorado-Boulder) was an excellent host and all the Conference was wonderful.

"I also had the opportunity to participate in a meeting organized in Argentina by José Miguel Abraham. This was a great meeting, with a small group of very concerned people, with a lot of proposals, so simple and so excellent!! Right now I am honored as being a member of the Board of Directors, and I really appreciate that!"

Adela directed the IV Jornadas Internacionales para la enseñanza preuniversitaria y Universitaria de la Química (4th International Day for Teaching Pre-university and university-level chemistry) in Mérida, Yucatan, Mexico in November 2005, attracting hundreds of participants from throughout Latin America, Europe and the United States. The ICUC was well-represented with PBK giving a plenary talk and several other members, including CMC-A, presenting papers.

At about the same time that we joined forces with Adela Castillejos Salazar and José Miguel Abraham, we had the good fortune to have JMA invite a very accomplished faculty member from Spain, Gabriel Pinto Cañón, (GPC) who brought to our organization a vast understanding of the European educational system during a time of great change, as the European Union was forming, and a common educational system, based on the Bologna Process (*51*), was being introduced formulated and implemented across many European countries.

IV. Gabriel Pinto Cañón

Born in Madrid (Spain); Male; White; 47 years old; and wife also from Madrid; two daughters (14 and 9 years old); public education and public University; dad is doctor pediatrician (now retired) and mom is housewife; chemist; chemistry educator; travel and history (specially about XX Century in Spain) fan.

One fine example of how ICUC has had a tremendous effect in the professional development of its members is the report that Gabriel Pinto Cañón sent to us in 2010 after six years of fruitful labor: "Although I knew CMC-A and PBK, I joined ICUC in 2003. Since that time, ICUC opened my horizons as chemical educator, exchanging ideas and experiences with teachers from many countries with special emphasis in first year courses of Chemistry."

"This exchange has been done mostly in a virtual environment (Yahoo Messenger) or via e-mails. However, I consider very important to mention that it was also very useful to have personal encounters in the conferences organized by ICUC : Urbana-Champaign (2005) and Boulder, Colorado (2007). In the first conference I was a keynote speaker and my conference was published in *The Chemical Educator* (*52*).

"With ICUC collaboration I organized several one-day meetings, named Jornadas, with the attendance of many Chemistry Teachers from many countries."

Two more recent ICUC members from Latin America have made important contributions to the organization. Patricia Morales Bueno, from the Pontifical Catholic University of Peru, in Lima, Peru joined ICUC in November of 2004. Now she is a member of the Board of Directors. Patricia is in charge of one of the recently created ICUC communities that will gather teachers with common interests worldwide. This community is "Problem Based Learning (Aprendizaje basado en problemas)" (*53*).

Amalia Torrealba, from Universidad Central de Venezuela, joined the ICUC in June 2006. She is the head mentor for the Venezuelan delegations of the International Chemistry Olympiads. One of her goals for 2010 is to bring more teachers from her country to join ICUC. Here are some of Amalia's comments about the future direction of the ICUC:

"Since I have been an ICUC member, my experience has been very interesting. The fortress of this association is the integration of a considerable number of teachers from several countries, which allow us to examine educational problems from different perspectives, and to generate ways to solve them. We have had the opportunity to share with colleagues in the meetings, like the 2nd FYI Conference in 2007, held in the University of Colorado-Boulder, USA.

"The creation of the ICUC Communities is a great idea to encourage the participation and the exchange of ideas between teachers from different places. It allows us to create networks of participation and collaboration on specific topics, and to know other teaching programs in Chemistry. Venezuela, as in other countries, is developing diverse programs related to experimental research, interdisciplinary sciences, and scientific projects in which problems of the communities are solved applying chemistry knowledge. The ICUC can create a database to do research compiling information of all countries, where plans are generated in cases of similar situations attending to cultural diversity and the curricula of every country. Venezuelan teachers are always asking for workshops to improve their teaching abilities; the ICUC might develop programs for this, and / or facilitate exchanges between teachers, a great incentive for them. Another interesting aspect to develop is also to do a network between chemistry students of different countries. For the year 2011, one of the activities that in Venezuela is to do a meeting with forums that generate documents about the Chemistry and its role in developing societies. The support of the ICUC would be fundamental in this activity."

Conclusion: The Challenges and Opportunities That Lie Ahead

The work done by many ICUC members in the last six years shows the benefits of having an international network that allows chemistry educators to share their knowledge and experience. The great challenge for the organization and its members will be to keep the momentum in the face of the economic recession that restricted travel, and made on-line methods of communication the single way of keeping in touch. We continue to work with our Ning-based communities, Yahoo-Messenger-based Board of Directors meetings, and the ubiquitous e-mail. A primary goal for the near-term is to be able to find ways to meet at conferences, because this personal contact gives the sense of support and friendship that is so important.

Even with the recent economic recession, the broad initiatives and specific activities developed under ICUC umbrella have had a profound impact on our everyday teaching work, and have assisted us intellectually, socially and, in more than a few cases, financially, with university incentives for publications, as well as career advancement.

In many Latin American and Iberian universities, chemists focused on research have an unquestioned professional advantage over those who dedicate ourselves mostly to teach chemistry; we reject this hierarchy of research over teaching, and believe our work with students to be of the greatest possible value as we deal with the great scientific and social challenges of our increasingly international world. The ICUC provides an international forum where we have been able to show that our teaching skills are valued. The belief in the power of education - of working with first-year students at this most important time in their lives to make a better world - is the common culture that we in the ICUC share.

References

1. Culture. *Merriam-Webster Online Dictionary*; 2009. http://www.merriam-webster.com/dictionary/culture (accessed March 10, 2010).
2. Mervis, J. Many voices, one message. *Science* **2007**, *317*, 63.
3. Stigler, J.; Hiebert, J. *The Teaching Gap*; The Free Press, Simon & Schuster: New York, 1999.
4. Intensive English Institute, University of Illinois at Urbana-Champaign. http://www.iei.uiuc.edu (accessed January 6, 2010).
5. The International Center for First-Year Undergraduate Chemistry Education (ICUC). http://icuc.unk.edu/english/eng_home.htm (accessed January 6, 2010).
6. Oshkosh Science Outreach, University of Wisconsin. http://www.uwosh.edu/science_outreach/ (accessed January 6, 2010).
7. Kelter, P. B. Razones por las que la enseñanza de las ciencias debe cambiar. *Educación Química* **1992**, 129.
8. Kelter, P. B.; Castro-Acuña, C. M. Estrategias para aumentar la quimicultura en nuestra sociedad. *Anuario Latinoamericano De Educación Química* **1996**, *2*, 308.

9. Kelter, P. B.; Carr, J. D.; Johnson, T.; Castro-Acuña, C. M. The chemical and educational appeal of the orange juice clock. *J. Chem. Educ.* **1996**, *73*, 1123. This paper was also the lead paper for the *J. Chem. Educ. Online*.
10. Kelter, P. B.; Castro-Acuña, C. M. Sugerencias para mantener actualizados nuestros cursos de quimíca. *Anuario Latinoamericano De Educación Química* **1997**, *IX, IX, II*, 26–33.
11. Neimoth-Anderson, J.; Carson, G. A.; Dormedy, D.; Emal, C.; Haas, E. J.; Johnson, T. L.; Mazlo, J.; Urlacher, T. M.; Kelter, P. B.; Llano, M.; Dominguez, R.; Castro-Acuña, C. M. Métodos motivacionales para el laboratorio de química general. Un estudio comparativo do universidades *primera* parte. *Anuario Latinoamericano De Educación Química* **1998**, *X, II*, 185–190.
12. Neimoth-Anderson, J.; Carson, G. A.; Dormedy, D.; Emal, C.; Haas, E. J.; Johnson, T. L.; Mazlo, J.; Urlacher, T. M.; Kelter, P. B.; Llano, M.; Dominguez, R.; Castro-Acuña, C. M. Métodos motivacionales para el laboratorio de química general. Un estudio comparativo do universidades *segunda* parte. *Anuario Latinoamericano De Educación Química* **1999**, *XI, I*, 294–300.
13. Dominguez, R.; Castro-Acuña, C. M.; Grundman, J.; Kelter, P. B. Puzzles in chemistry and logic. *J. Chem Educ.* **1999**, *76* (4), 496–498.
14. Kelter, P. B.; Castro-Acuña, C. M. ¿Quién debe enseñarles ?: El extraño caso del aspecto económico contra los estudiantes de química general. *Anuario Latinoamericano De Educación Química* **2001**, *XIV, XII*, 153–156.
15. Kelter, P. B.; Castro-Acuña, C. M. The National SAT exam: Its use and misuse as a ticket to higher education. *Anuario Latinoamericano De Educación Química* **2002**, *XV*, 304–308.
16. Castro-Acuña, C. M.; Abraham, J. M.; Kelter, P. B. *Enfoque De Química Cotidiana En El Anuario Latinoamericano De Educación Química.* Proceedings of the Jornada sobre Didáctica de la Química y Vida Cotidiana; Pinto, G. C., Ed.; Madrid, Spain, May 29, 2003.
17. Abraham, J. M.; Castro-Acuña, C. M.; Kelter, P. La función de la investigación educativa, científica y tecnológica vinculadas ante la necesidad de un desarrollo sustentable. Proyectos educativos integrales (PEI). *Anuario Latinoamericano De Educación Química* **2004**, *XVII, XVIII*, 91–98.
18. Abraham, J. M.; Castro-Acuña, C. M.; Kelter, P. Hacia una educación integral en ciencia y tecnología. Su aporte al desarrollo sustentable. *Anuario Latinoamericano De Educación Química* **2004**, *XVII, XVII*, 63–68.
19. Middlecamp, C.; Kelter, P. B.; Castro-Acuña, C. M. First-year chemistry (arguably). *Chem. Educ.* **2004**, *9*, 182–183. Also published in Spanish, as "Química de Primer Año (Discutible)," in the same journal.
20. Kelter, P. B.; Castro-Acuña, C. M.; Middlecamp, C. The case for an international center for first-year undergraduate chemistry education as a means of dealing with a mid-life crisis. *Chem. Educ.* **2004**, *9*, 255–256. Also published in Spanish, as "Como un Centro Internacional para la Enseñanza de la Química en el Primer Año de Educación Superior puede contribuir a superar una crisis como la de los 40 años" in the same journal.

21. Abraham, J. M.; Castro-Acuña, C. M.; Domínguez-Danache, R.; Pinto, G.; Kelter, P. Chemical education and sustainable development: Recommendations of the International ICUC-PIEQ XV Conference. *Chem. Educ.* **2007**, *12*, 292–294.
22. Abraham, J. M.; Castro Acuña, C. M.; Kelter, P. La investigación científica, tecnológica y educativa como alternativa para el desarrollo sustentable y contribución a la superación de la situación social, económica y ambiental. Perspectiva: proyectos educativos integrales (PEI). *Anuario Latinoamericano De Educación Química* **2009**, *XXIV*, 93–97.
23. Castro Acuña, C. M.; Kelter, P. B. *Presence of Industry in Mexico's Chemistry Curriculum.* American Chemical Society National Meeting, Anaheim, California, April 1995.
24. Castro-Acuña, C. M.; Domiguez-Danache, R.; Kelter, P. B.; Grundman, J. *Problem Solving with Logical Chemistry.* National Science Teachers Association Annual Conference, Las Vegas, NV, April 1998.
25. Castro-Acuña, C. M.; Domínguez-Danache, R.; Kelter, P. B.; Grundman, J. *Logical Chemistry: A Different Challenge for Your Students.* American Chemical Society National Meeting, Dallas, TX, April 1998.
26. Dominguez, R.; Castro-Acuña, C. M.; Kelter, P. B. *Learning Science with Logic Puzzles.* National Science Teachers Association National Conference, Boston, MA, March 1999.
27. Kelter, P. B.; Castro-Acuña, C. M.; Dominguez, R. *Teaching Science in Context with Social Values.* National Science Teachers Association National Convention, Orlando, FL, April 2000.
28. Castro-Acuña, C. M.; Kelter, P. B.; Dominguez, R. *Teaching Equilibrium with a Worksheet.* National Science Teachers Association National Convention, Orlando, FL, April 2000.
29. Castro-Acuña, C. M.; Domínguez, R.; Kelter, P. B. *Uso de Problemas de Razonamiento Lógico para Reforzar la Enseñanza Química.* XIX Congreso Nacional de Educación Química, San Luis Potosí, México, September 2000.
30. Kelter, P. B.; Castro-Acuña, C. M.; Dominguez-Danache, R. *One Philosophy in Two Countries: Common Ground, Common Programs between Long-Time Friends.* 221st American Chemical Society National Convention, San Diego, April 2001.
31. Castro-Acuña, C. M.; Domínguez, R.; Kelter, P. B. *Logical Chemistry for High School Students.* Chem Ed 2001, Toronto, August 2001.
32. Domínguez, R. Castro-Acuña, C. M.; Kelter, P. B. *Learning More from Your Mistakes than from Your Successes.* Chem Ed 2001, Toronto, August 2001.
33. Castro-Acuña, C. M.; Domínguez,R.; Kelter, P. B. *Enseñanza de la química mediante problemas de razonamiento.* XX Congreso Nacional de Educación Química Ixtapa, Gro., México, September 2001.
34. Kelter, P. B., Vestal, J.; Castro-Acuña, C. M.; Domínguez-Danache, R. *Análisis Comparativo de Exámenes Diagnóstico Utilizados en los Estados Unidos de America y México.* Sociedad Química de México, National Conference, Cancun, México, September 2002.
35. Castro-Acuña, C. M.; Domínguez, R.; Kelter, P. B. *Diseño De Problemas De Razonamiento Lógico Para Reforzar El Aprendizaje De La Química En*

Diferentes Niveles De Enseñanza Sociedad Química de México. National Conference, Cancun, Mexico, September 2002.
36. Kelter, P. B.; Walsh, J. L.; Castro-Acuña, C. M. *Principles of Electrochemistry via the Orange Juice Clock and Other Demonstrations.* 225th American Chemical Society Annual Convention, New Orleans, March 2003.
37. Abraham, J. M.; Castro-Acuña, C. M.; Kelter, P. B. *Las Relaciones Complejas Entre El Desarrollo Socio-económico-ambiental Y La Educación Científica (Particularmente Química) Y Tecnológica".* VI Jornadas Nacionales y III Internacionales de Enseñanza Universitaria de la Química, La Plata, Argentina, September 2003.
38. Kelter, P. B.; Castro-Acuña, C. M.; Abraham, J. M. *Collaboration as the Underlying Feature for Advancing the Agenda of Chemical Educators: International Center for First-Year Undergraduate Chemistry.* 18th BCCE, Ames, IA, July 2004.
39. Castro-Acuña, C. M.; Abraham, J. M.; Kelter, P. B. *Integrated Educational Projects: Teaching Chemistry in Connection with Personal and Social Development.* 18th BCCE, Ames, IA, July 2004.
40. Castro-Acuña, C. M.; Kelter, P. B.; Pinto Cañón, G.; Abraham, J. M. *Integrated Educational Projects: Teaching Chemistry.* 229th American Chemical Society National Meeting, San Diego, CA, March 2005.
41. Kelter, P. B.; Castro-Acuña, C. M. *How the Dignity of Teaching Provides a 13-Year Framework for Change in Chemical Education.* 229th American Chemical Society National Meeting, San Diego, CA, March 2005.
42. Castro-Acuña, C. M.; Kelter, P. B. *Centro Internacional para la Enseñanza de la Química.* XXIV Congreso Nacional de Educación Química, Morelia, Michoacán, México, September 2005.
43. Abraham, J. M.; Castro-Acuña, C. M.; Kelter, P. *Aspectos principales a considerar en el diseño de programas para la enseñanza de la ciencia (particularmente química y tecnología), útiles para una opción de desarrollo sustentable. Contexto: proyectos educativos integrales (PEI).* IV Jornadas Internacionales para la Enseñanza Preuniversitaria y Universitaria de la Química, Mérida, Yucatán, México, November 2005.
44. Kelter, P. B.; Castro-Acuña, C. M.; Abraham, J. M. *The Importance of International Collaboration in Chemical Education – Introducing The ICUC.* IV Jornadas Internacionales para la Enseñanza Preuniversitaria y Universitaria de la Química, Mérida, Yucatán, México, November 2005.
45. Castro-Acuña, C. M.; Kelter, P. B. *The Importance of International Collaboration in Chemical Education via the ICUC.* National Science Teachers Association Annual Conference, Boston, MA, March 2008.
46. Abraham, J. M.; Castro-Acuña, C. M.; Kelter, P. B. *Education in Science and Technology Based on Integrated Educational Projects.* Bicentennial Conference on Chemical Education, Bloomington, IN, July 2008.
47. Castro-Acuña, C. M.; Kelter, P. B. *The Importance of International Collaboration in Chemical Education via the ICUC.* National Science Teachers Association International Science Day, New Orleans, LA, March 2009.

48. ICUC Newsletters. The International Center for First-Year Undergraduate Chemistry Education. http://icuc.unk.edu/english/eng_newsletters.htm (accessed January 7, 2010.)
49. The Ning Social Networking Main Site. http://www.ning.com/ (accessed February 2, 2010).
50. Abraham, J. M.; Azar, M. L.; Segovia, R. F.; Castro-Acuña, C. M. Integrated educational projects: A teaching technique designed to favor the overall development of our students. *Chem 13 News* **2000**, *287*, 7.
51. The European Union Description of the Bologna Process. http://ec.europa.eu/education/higher-education/doc1290_en.htm (accessed February 3, 2010).
52. Pinto, G. New general trends of university (including first-year chemistry) education in Europe. *Chem. Educ.* **2005**, *10*, 317–319.
53. Problem-Based Learning. Ning Web Site. http://icucabp.ning.com/ (accessed March 6, 2010).

Chapter 9

Emerging Pharmaceutical Infrastructure in Sub-Saharan Africa

Rolande Hodel*

AIDSfreeAFRICA, 125 S. Highland Ave. # 3-B1, Ossining, NY 10562
***E-mail: RRHodel@aol.com. Phone: 914-923-2073.**
Web site: www.AIDSfreeAFRICA.org.

AIDSfreeAFRICA is a non-profit organization dedicated to establishing sustainable drug production in Sub-Saharan Africa. This chapter discusses how the organization was founded and the initial steps made towards establishing pharmaceutical production in Cameroon. This chapter also indicates how lack of infrastructure and resources has limited the expansion of such facilities. The educational system in Cameroon is discussed along with barriers to the training of industrial chemists. The chapter concludes with the outlook for AIDSfreeAFRICA in Cameroon.

Introduction

From global warming to global terrorism, today we know that what happens in a remote corner of the world can affect us all. In Africa, where the vast majority of the population lives on less than a dollar per day, desperation may lead to desperate actions, with catastrophic results. What happens in Africa matters to the rest of the world. There is general agreement that improving overall health is a necessary prerequisite to lifting a country out of poverty. African governments already spend a considerable portion of their gross domestic product on health care for their citizens. However, according to World Health Organization estimates, 270 million people in Africa lack access to even the most essential medicines (*1*). This "lack of access to medicines…kills an estimated 6 million people every year, mostly in Sub-Saharan Africa." (*2*) My experiences in Africa highlight the paramount role that pharmaceutical production in Africa will have to play to reaching the goal of achieving universal access to medicine and to facilitate economic development.

The Melinda and Bill Gates foundation supported the Grand Challenges in Global Health initiative with US$450 million, believing that science can and will provide solutions for the world's most pressing health issues (*3*, *4*). For global issues such as global warming, terrorism, health, tsunamis, and asteroids, to name a few, scientists will find solutions. Today we take breathing smoke-free air for granted, thanks to catalytic converters and lead-free gasoline. The world is now free of the scourge of Smallpox. Polio will soon be eradicated as well. We do not know in advance just what a solution will look like, but solutions are found every day.

To better understand the impact of globalization on industry, and the need for chemistry programs to educate scientists better suited to fill the resulting new positions, I will point out obstacles encountered during the past five years in my efforts to advance local drug production in the African country of Cameroon. This will be followed by some historical facts and a description of the educational system. Discussion of my multicultural work in economically impoverished regions will hopefully provide insights into what it means to teach and organize in an environment with spare resources. After describing day-to-day challenges in drug production, limits to education, and the unique opportunities and factors that are present that offer real hope, I hope to convey the importance and tenability that drug production is possible in Africa.

The Founding of AIDSfreeAFRICA

AIDSfreeAFRICA was conceived in 2003 as the first non-profit organization that would advocate for, and implement drug production in Sub-Saharan Africa. Efforts to develop the infrastructure for medical care, and importation of the drugs necessary to provide care, however, have been ongoing in this region. From the beginning of Christianization, the local Churches in Africa have been successful in establishing hospitals that provide quality care in addition to, or sometimes exceeding, the care provided by government-owned and managed district hospitals. However, African hospitals lack access to a sufficient quantity of quality drugs and medical supplies. Until an infrastructure is created that supports in-country drug production, this problem will persist.

My experiences with AIDSfreeAFRICA suggest that while the Clinton Foundation, Doctors Without Borders, the Bill and Melinda Gates Foundation, and numerous other philanthropic organizations have made persistent efforts to provide much needed medicines in a timely manner in continents such as Africa, the sheer volumes of medicine needed in a sustained way, along with the need to reach remote regions, has exceeded the capacity of these organizations to provide the quantity of drugs needed to make a significant difference in the lives of the people in need.

In Cameroon, where my efforts have been focused, pharmacists who struggle day in and day out to provide the hospitals with imported drugs have long been thinking about how to establish local drug production. Thus, in Africa, hospital care and drug production go hand in hand. In most western countries, doctors know drugs are available and they prescribe whatever they feel the patient needs without even thinking of the possibility that the drug may not be in the pharmacy. In Cameroon, for example, hospitals are plagued by stock outs (empty warehouses), so drug availability is always on the doctor's mind. If a drug is produced far away, it cannot be made available on short notice. Life-saving drugs can also be withheld for political reasons. If the drugs are produced locally, the drugs can be made available on short notice and government bureaucracy may be circumvented by private industry initiatives. On a visit to Cameroon, the first thing doctors told me was: "Please, Doctor, bring us drugs but make sure they are continuously available." In the US we do not appreciate how much it means to have drugs available on demand.

In the process of founding AIDSfreeAFRICA, I studied successful organizations and how their structure and modus operandi allowed them to be successful. I was looking for role models and events in history to guide me. I chose "The Hunger Project," a twenty five year old established non-profit organization working around the world with the goal of eradicating hunger, as a good model. The organization's roots are firmly anchored in the teaching philosophy of Landmark Education Corporation, with which education I am familiar. I adopted the visionary style of Landmark, and combined it with the "hands-on, boots-on-the-ground" practical approach of The Hunger Project.

In 2003 I took the opportunity to represent Servas, an international peace organization at the United Nations Economic and Social Council (ECOSOC). I had been a member since 1989. In my capacity as representative of Servas I attended an international conference where President Lula da Silver gave the keynote address. He told us how drug production of antiretrovirals had turned the tide and stemmed the pandemic in Brazil by making the costs of drug production and distribution affordable to the government, allowing them to provide the drugs at no cost to many infected individuals. Sitting in the audience in 2003, I thought: "This is cool; we will do this in Africa!"

While the story of Africa has yet to be written, Brazil has demonstrated how readily available drugs can have a huge effect on citizens' health. In 1990, with over 10,000 Brazilians infected with HIV, the World Bank predicted that, by the year 2000, 1.2 million people in Brazil would be living with HIV (*5*). Owing to the development of improved drugs by 1996, now used as cocktails that could simultaneously attack different stages of the HIV lifecycle (designated highly active antiretroviral therapy; HAART) (*6, 7*) combined with the effectiveness of stepped-up prevention campaigns (comprised of drug production, distribution, and education) in conjunction with numerous non-government organizations (NGOs), the actual number of individuals living with HIV in 2000 was estimated to be 600,000, half of what had been predicted by the World Bank (*3*). In contrast, HIV infection was spreading exponentially in Africa and the Caribbean where no such campaigns were in place. This bold program in Brazil illustrates how readily and powerfully available drugs can impact health outcomes.

In monetary terms, the Ministry of Health estimated in 2002 that the availability of antiretrovirals had prevented approximately 358,000 HIV-related hospitalizations, saving more than US$1.1 billion." (*8*) By the end of 2007, it was estimated that 80% of those requiring the drugs were receiving them (*7*). According to the World Health Organization (WHO), this level of treatment coverage is more representative of that of a more developed nation, owing in great part to Brazil's ability to produce several AIDS drugs locally. By 2006, Brazil's large pharmaceutical industry was producing approximately 40% of its ARVs (*9*).

Since 1996, Brazil has complied with the international Agreement on Trade Related Aspects of Intellectual Property (TRIPS), which was established to protect the patent rights of pharmaceutical companies. A major tool in these negotiations was a clause in the TRIPS agreement that allows developing countries to issue 'compulsory licenses' for drugs. Compulsory licenses allow countries to override patent laws and produce their own generic (copied) versions of company-owned drugs, and can be issued when the government of a developing country deems it to be a public health emergency (*10*).

Brazil demonstrated that local drug production makes a dramatic difference. But Brazil is hardly thought of as developing country and it already had local drug production capacity. For Brazil, it became crucial to produce antiretrovirals whose patent protection had to be broken due to the nation's health crisis. This made it monetarily possible to provide these drugs to so many of those in need.

Given the realization that this kind of approach is likely to be needed in other nations, including those far less developed than Brazil, it is critical to ask if there are precedents for establishing drug production in developing nations. Indeed, only fifty years ago, Chemical, Industrial and Pharmaceutical Laboratories, Ltd. started operation of a plant in Vikhroli, Mumbai. Popularly known by the acronym CIPLA, today, the company exports drugs to 180 countries world-wide (*11*). In 2005, India joined the World Trade Organization and agreed to adhere to international patent and trade agreements (TRIPS). According to AIDSfreeAFRICA collaborators in Cameroon, an estimated 85% of the antiretrovirals drugs available in Sub-Saharan Africa are generics delivered by CIPLA. According to doctors and medical professionals in Cameroon, today, the majority of drugs they have problems importing are generics which are NOT in the category of antiretrovirals. India, a developing country with a population of over a billion, has developed two of the world's four largest generic drug manufacturing companies, namely CIPLA and Ranbaxy. Of the other two, GlaxoSmithKline also exports drugs to Africa, while Teva (Israel) as is the world largest generic drug manufacturer focuses more on Latin America and has only a small share of the African market.

It is a pity that despite all efforts by non-profit organizations and commercial enterprises the current total import of drugs into Africa falls an estimated 50% short of what is urgently needed (*12*). Certainly, there is a need and a market for quality drugs. More importantly, my experience with AIDSfreeAFRICA indicates that drugs do not have to be donated to fill this gap; they can and need to be sold. Principally, there are four ways drugs come into Cameroon and to the patient:

- First, foreign non-profit organizations raise money, buy drugs, import them into African countries and distribute them for free. This approach involves no licensing requirements, no taxes and no government interference. The disadvantage to this approach is that its lack of regulation invites abuse. Organizations often close their eyes to the fact that these donated drugs are frequently *sold* to patients, and not given to them as they were intended.

- Second, drugs are imported and bought through a distribution agency managed by the government and by private wholesalers. The government requires that all drugs have to be registered by a cumbersome system similar to that used by the United States Food and Drug Administration for new drug approvals. This does not work within the context of African governments and creates huge backlog and long waiting time. It inappropriately treats applications for generic essential drug licenses with the same rigor as new drug applications. In addition, imports of drugs are limited by the size of the government budget.
- Third, drugs are obtained from a flourishing black market that operates unchecked. While black market drugs fill a desperate need, they are unregulated and lacking in any traceable history or guarantee (with many of these drugs coming from Nigeria). Thus, their chemical composition, dosage, impurities, and treatment in transport are all unobtainable. More recent developments are further complicating the situation by confusingly equating generic drugs with counterfeit drugs.
- Fourth, drugs can be produced locally. The advantage to local production is that this can close the gaps in the availability of drugs, and in their continuity, as seen in the example of Brazil. The disadvantages right now are that Cameroon, and many other underdeveloped nations, lack not only the infrastructure for production, but also appropriate government regulation and means of drug distribution. Locally produced drugs, as we envision it, will be registered with the government, and sold to local wholesalers and hospitals. Production and facilities will be quality controlled through appropriate government agencies and eventually WHO approved.

As mentioned before, fifty years ago there was no pharmaceutical drug production in India, but today the Indian company CIPLA and others provide the bulk of the pharmaceuticals consumed in Sub-Saharan Africa. It is time for Africa to build its own pharmaceutical production capacity. For this purpose, I visited Kenya and Cameroon in 2005 to conduct needs assessments and feasibility studies for the founding of in-country drug production facilities. Since then, AIDSfreeAFRICA has focused on Cameroon.

Background and History of Cameroon

AIDSfreeAFRICA has been working in Cameroon for the past five years. When I present my work to an audience someone will inevitably ask, "Why Cameroon?" It happened by chance, and it turned out to be a good choice. I had been asking friends if they knew someone who worked in Africa who could make a recommendation. I was visiting friends in Kansas when Dr. Gunda Georg, who was at the University of Kansas, Department of Medicinal Chemistry, recommended her friend Dr. Carole McArthur, a Professor at the University of

Missouri, Kansas City, who, according to Dr. Georg "was working somewhere in Africa." As it turned out, Dr. McArthur has an on going research collaboration with a Cameroonian, Dr. Paul Achu, at the Mezam Polyclinic in Bamenda, Cameroon. I presented my idea, and the rest is history.

I found Cameroon to be a microcosm of Sub-Saharan Africa. A midsize country, it is somewhat larger than California, with almost twenty million inhabitants http://www.maplandia.com/cameroon/. Culturally it belongs to West Africa, but economically it belongs to Central Africa. It occupies the right angle on Africa's western coast, and stretches from the Saharan dessert in the north to the equator in the south where it has 400 km (216 miles) of coast line, rainforest, savanna, volcanic ranges flecked with crater lakes, and the highest mountain on the west side of the continent. Mt. Cameroon stands 4095 meters tall (13,435 feet), rising directly from the ocean shore. The country is 16% Muslim, 33% Christian, with the rest holding native tribal beliefs. Officially bilingual in French (government) and English (North West and South West Region), and most people speak Pidgin English and one of 270 distinct native tribal languages. President Paul Biya, in power for 29 years, and not willing to hand over the reins, has been only the second President since the country's independence from the French on January 1, 1960. He succeeded Ahmadou Ahidjo, who had held power for 22 years, in 1982. Biya recently changed the country's constitution, allowing him to run again for president in 2011. The result of his reelection is likely to be a foregone conclusion for the many disappointed citizens who yearn for a change in government.

Cameroon is a stable and peaceful country which has never had a civil war. The government spends 3% of the gross national income (GNI) on health, 12% on education and 10% on defense according to UNICEF data from 2007 (*13*). However, the population has much to complain about. The gross national income (GNI) per capita is listed as US$1,150. One third of the population has a daily income of US$1.25/day or less, well below the US$29.67/day of the U.S. poverty line. For women the lifetime risk of maternal death due to complications from childbirth is a staggering 1-in-24. Valued strongly by Cameroonians, primary school attendance is 86% for boys and 81% for girls. High school attendance drops significantly for both genders to 28% for boys and 22% for girls. The gender gap becomes more obvious when looking at the 72% literacy of 15 to 24 year old boys, *versus* 59% for the same age female population. Teenage girls can be found everywhere as unpaid domestic workers, working for room and board only (*13*). Boys on the other hand are glad if they manage to find work driving a car or motorbike as a taxi.

The agricultural sector employs three-quarters of the work force, and contributes approximately half of the gross domestic product. Favorable agricultural conditions, such as rich volcanic soil and plenty of water in conjunction with landownership by local families, would suggest a level of survival incommensurate with the poverty levels observed. Cameroonians tell me that the country is so peaceful because there is always food available. Since people have plenty to eat they don't fight the government. Although this may be true for most of the country, it changes dramatically traveling north is where one encounters desertification and hunger.

Cameroon exports energy generated from hydro-electric power and is currently erecting its first off-shore oil drilling platform. The country mines precious metals and exports timber and rubber. So far, Cameroon has not taken advantage of the potential to make money through tourism, but there is promise for the future.

The chemical industry is small. There is the French-owned company Air Liquide in Douala which sells compressed gas in cylinders. A few international chemical and pharmaceutical companies keep marketing and sales offices. Chemicals are generally imported through Nigeria. Chinese goods flood the local markets everywhere, endangering fragile local production and craftsmanship. European beer breweries are plentiful -- with highly secured, fenced-in production facilities that are well managed, and fleets of in-house trucks rivaling all others in the country. Cameroon is rumored to be close to Ireland and Germany in per capita beer consumption even though 660 ml (22.3 ounce) bottle of delicious beer costs 500 CFA (the equivalent US$1.00), which represents the average daily income for roughly 1/3 of the country's inhabitants!

Despite this, I was delighted when I saw my first billboard depicting the vats of beer, copper piping, and gowned workers brewing beer. I reasoned that since Cameroonians can brew beer, they can be trained to produce pharmaceuticals. After all, beer production must be clean and controlled, needs purified water and handling of products for human consumption. The goods must be processed, packaged, labeled, shipped and monitored. The facilities must be maintained and repaired. Security must be in place, with the raw materials imported and stored safely and under controlled conditions. The billboard illustrated that there are people who have been trained to perform and manage all of these tasks. In fact, the first person we hired was an accountant who had formerly worked for one of the beer breweries.

To this day, I believe Cameroon was a good choice for the country in which to set out our mission. The most notable advantages are the stability of the country and the peace-loving nature of the people. There is little violent crime. The fact that there is plenty of food is always cited as the reason for the peaceful nature of the otherwise tribal and inhomogeneous population. Cameroon is not on the United Nations list of the fifty poorest countries. Thus, as weak as the infrastructure is, there are plenty enough resources to work with and there are educated people eager to learn and collaborate.

First Steps towards Pharmaceutical Drug Production

Since 2005, I have spent a total of thirteen months in Cameroon. During my first two-month trip to Cameroon, my chief focus was to establish needs assessments. I needed to learn as much as possible and to assess what there was in the country to build on, what was missing, what was needed, and where the bottlenecks were with respect to drug production and access to drugs. Perhaps the most surprising finding was that *access to antiretrovirals was not the biggest problem*. In fact, it was not even a problem, since these drugs were provided and paid for by the Global Fund. As a consequence, AIDSfreeAFRICA shifted its focus from producing antiretrovirals to what was needed the most: *essential generic drugs* for the most common maladies.

But first, more about my host Dr. Paul Achu, member of a large and well respected family in Cameroon. His oldest brother Simon Achu had been the Prime Minister of Cameroon. Paul had standing in the Society which allowed him to introduce me in a most proper and useful manner. I am grateful for these advantageous introductions. I also had the opportunity to learn the basics. Amongst other things I learned how people live and do business in Cameroon. Dr. Achu is a British educated pathologist who is today the proprietor of the Mezam Polyclinic, an approved HIV/AIDS treatment facility in Bamenda, in the North West region of Cameroon. I have had the opportunity to meet with doctors, pharmacists, nurses, social workers, business men and women, politicians, professors, students, storeowners, women selling goods at the market, and various others -- too many to count.

Focused as I was on pharmaceutical production, I met and assessed possible local collaborators. It was never the intent of AIDSfreeAFRICA to build a factory ourselves. Instead, we have wanted to find Cameroonians interested in collaborating with us so that we could support *their* endeavors!

During the first three years of AIDSfreeAFRICA the work focused on one start-up company, namely Diamond Pharmaceuticals which was located in Douala, the largest city in Cameroon. I was repeatedly asked if the company would survive and what I would do were it to fold. Statistics from the Small Business Administration (SBA) show that "two-thirds of new employer establishments in the USA survive at lease two years, and 44 percent survive at least four years." This confirms what my business coach from "SCORE" (Senior Core of Retired Executives) told me, *i.e.*, that in the USA, 60% of all business start-ups fail (*14*). Thus, my answer was and is, “If 60% of start-up businesses in the US don’t make it past four years, why would I expect 100% success in Cameroon?” So far, neither Diamond Pharmaceuticals nor Meditech, both start-up companies for which we consulted, managed to survive. Determined to learn from these experiences I am convinced it has brought us step closer to finding solutions.

In my opinion, it will take a critical mass of several companies to achieve the overall goal of establishing a pharmaceutical industry in Sub-Saharan Africa. After that, it will be easier for others to follow in their footsteps. Attracting the interest of the Cameroonian Prime Minister Philemon Yang, and the US Embassy Business Development Services in Cameroon, has proved very helpful in making connections with business people and Cameroonian investors. We looked for more established groups and were introduced to the head of the Cameroonian Baptist Convention, Professor Pius Tih. In addition to operating 5 large hospitals and 23 Health Clinics, the Cameroonian Baptist Convention operates a production site in Mutengene south of Douala (*15*). This site produces sterile intravenous fluids, eye drops and ointments. However, their non-profit church status limits them to in-house production and in-house distribution of these products. In other words, the Baptist hierarchy, in order to avoid government requirements such as taxation, inspections and licensing, made the choice to limit production for distribution within their own hospitals.

Currently, Genemark an established company in Douala has become our main focus. The company is family owned, in operation for more than five years and expanding steadily. They are the only company in the country to produce children's medicines in the form of syrups, mainly quinine (for treating drug resistant malaria), paracetamol (Tylenol) and carbocysteine (a cough syrup). They have recently expanded into producing vitamins, especially the urgently needed iron/B12 vitamin combination to address the severe iron deficiency of Cameroonian women. The iron supplement is usually formulated with vitamin B12 since processes involving blood generation need both, iron and B12. When one realizes that women who test positive for the HIV virus cannot be placed on antiretroviral treatment if their blood iron level is below normal, one can appreciate the consequences of iron deficiency. In the absence of iron supplements, doctors resort to giving iron-deficient women blood transfusions. Given the lack of blood banks, the standard source of blood for transfusions is from an insufficiently tested relative, who the patient is required to bring to the hospital, increasing the threat of contracting HIV and highlighting the importance of making iron/vitamin supplements available.

It can be argued that producing syrups is pharmaceutical production and I would not object. However, AIDSfreeAFRICA's gold standard for pharmaceutical production has been to push for production of *blister-packed solid oral tablets*. Liquids such as intravenous or IV fluids or syrups have been produced in Cameroon for some time. However, their use is limited. Most medicine needed is dispensed in tablet form and blister packaging protects drugs from heat, humidity and breakage. Thus we are very proud that we are now in the process of financing the purchase of blister packaging/labeling equipment. When this is accomplished, Genemark will have become the first factory to produce solid oral tablets of the malaria drug quinine, to treat the number one killer in Africa: malaria.

Another possible source for drug production in Cameroon is the ISO9001-approved formulations and packaging outfit, Kakwa Biopharm, located south of Douala (*16*). To this day, Cameroonian shareholders who live and work in the US have invested UD$ 3,000,000 and thus own the company. The goal has been to produce malaria drugs based on the WHO approved artesunate/amodiaquine combination drug. However, production has yet to begin. AIDSfreeAFRICA has been looking for US investments and is currently talking to a company in California. This company approached me asking for possible collaboration to bring their malaria drug on the market. Today the company is considering moving production to Cameroon. Hopefully, with these investment partners, the site can be validated and made ready to produce the generic drugs needed so desperately by Cameroonians. AIDSfreeAFRICA's role is to offer the US based investor's independent verification and long term oversight that is needed to bridge the gap in work culture and customs that may lead to a successful outcome.

On a recent trip to Cameroon, I had the opportunity to visit another possible center for drug production in Cameroon. Thanks to the US Embassy, the US Export Assistant Centers Gold Key Service, I had the opportunity to visit and tour of the production facilities of Cinpharm, in Douala. Cinpharm has ambitious plans for drug production, but as of today has not started production. Described as a state-of-the-art facility with stand alone source of purified water, electricity and clean air production, they are living up to the international standards for modern-day drug production. Their quality control-laboratories feature analytical equipment commonplace in the US but never seen in Cameroon, such as instruments for high pressure liquid chromatography and mass spectrometry. My first impression was that my job was done, as they had the capacity to produce drugs that could supply more than the needs for Cameroon. However, AIDSfreeAFRICA advisor William F. Haddad, Chairman and CEO of Biogenerics, Inc. (Brewster, NY) pointed out a few of the challenges Cinpharm was bound to face.

Cinpharm had done what I advise international investors to avoid, namely to build a company with features that do not fit Cameroonian reality. For example, although Cameroon is an exporter of energy, it is a very expensive commodity. Cinpharm was built for a country like India where energy costs much less. Furthermore, the equipment is high-tech, not easy to operate, and difficult to maintain and repair. Few people in Cameroon, if anyone, have the experience running and maintaining such complex production machines, particularly as such training is not provided in university settings. It will be a great challenge to maintain the current standards and to operate the facility in such an infrastructure poor country. Considering that it took three years for one hundred employees to be trained in these complex matters, and according to their chief pharmacist the company needs to hire and train three hundred more employees before running at full capacity, there is more work ahead. With the help from CIPLA and a six million euro loan from the German Investment and Development Association (DEG), they have managed to place a "Rolls Royce in the dessert". We hope the sand will not bring the engine to a grinding halt. AIDSfreeAFRICA is ready to do what it takes to see Cinpharm succeed. They have made a huge investment, ripe for the arrival of some experienced scientists willing to make the trip to Cameroon and to apply their expertise, to be generously rewarded by an experience only a third world country can offer.

As our mission states, we are interested in providing what is necessary for Cameroonians to produce affordable medicine. I believe that the best way to accomplish this goal is to help Cameroonians start and/or continue the building and expansion of pharmaceutical companies that can affordably produce medicine in the country. Now that we are convinced that production is inevitable, our focus has expanded to determining where these drugs will need to go and how they will succeed in getting there. According to the Minister of Public Health, 45% of the drug market in Cameroon is limited to the two largest cities, Douala and Yaoundé. Distribution beyond these two cities is severely hampered by road conditions and the lack of capital for purchasing drugs.

Through my involvement with the United Nations, I became familiar with the concept of the revolving drug fund, a concept endorsed by the United Nations. The concept is simple; a non-profit organization provides the capital needed to buy enough drugs to stock a hospital's pharmacy. The drugs are then sold to the patients and a portion of this money is paid back to the non-profit, which spends that money to buy more. In the absence of this commitment, it is frequently too tempting to use any accumulated amount of cash for one or another emergency. My collaborators in Limbe, with whom we established our first revolving drug fund, were not able to accumulate US$200 without misappropriating some money. Now we require daily deposits of smaller payments. The income generated through the revolving drug fund is used in part by AIDSfreeAFRICA to buy more drugs and supplies, thus ensuring continuous access to drugs and the hospital's ability to treat patients. The additional income generated has been used to pay salaries owed to nurses and for paying back-rent. In the future, this additional income can be used to expand the hospital service. Concomitant with these changes, patient attendance has increased ten-fold. This constitutes a favorable situation.

Although I personally raised over US$100,000 in cash donations and five -fold in in-kind services, it became obvious that our ability to finance our goals through donations only was not realistic since these donations were covering only small projects and the day-to-day running expenses of the organization. The next step for AIDSfreeAFRICA was to finance production equipment that runs into the hundred thousands. In 2009, as the economy continued to decline, we began looking for investors both in the US and in Cameroon. We also approached the pharmaceutical industry here in the US. Our first inquiries were met with the suggestion that we should start drug production, and that this would then generate interest in companies to invest. Well, this was putting the cart in front of the horse. We needed the money to start production, but were asked to start production in order to receive money. I did not give up.

In 2008, the Swiss-based pharmaceutical giant, Hoffman-La Roche, had offered us the possibility of a technology transfer to spin-coat an AIDS drug in Cameroon. Roche scientist Luc Schnitzler presented the program to our collaborators at the Cameroonian Baptist Convention production site in Mutengene. He then inspected the facility. Unfortunately, the Baptist facility was not in accordance with international production standards and lacked the space and necessary machines to spin coat an AIDS drug. The project has not progressed because Roche expects the technology transfer to proceed without them making any financial investments into upgrading a facility or to purchase machines and quality control equipment. Having visited Cinpharm in Douala, however, it is my intention to approach Roche again, particularly as Cinpharm has been approved by international standards and already has the spin-coating equipment. Unfortunately, Roche responded by informing me that this initiative has been completed.

While AIDSfreeAFRICA's main focus is on local drug production, we have also explored the benefits of importing and selling drugs that are produced elsewhere. I have forged a valuable interaction with the Belgian company Tibotec, a subsidiary of the US-based Johnson & Johnson, in bringing the much needed antifungal medication, Miconazole, to Cameroon. Tibotec offered us the drug and financing for registration and marketing. The project got us in communication with government agencies and gave us experience with drug licensing procedures, both of which will be valuable when drugs are produced locally. To date, we have succeeded in obtaining a license from the Ministry of Public Health, Department of Pharmacy, to import and sell the antifungal Miconazole nitrate 10 mg, muco-adhesive tablet. Miconazole is an imidazole antifungal agent, developed by Janssen Pharmaceutical. It is commonly applied topically, to the skin or mucous membranes, to cure fungal infections. It works by inhibiting the synthesis of ergosterol, a critical component of fungal cell membranes. Miconazole is used externally for the treatment of athlete's foot, ringworm and jock itch, and internally for oral or vaginal thrush (yeast infection). Depending on the application, miconazole is prescribed as an oral gel, topical cream 2%, pessary (vaginal deposit), or vaginal cream. Most of the miconazole products are available in cost efficient generic versions (*17*). The miconazole we import has been specifically developed as an easy to use oral adhesive that withstands the harsh climate conditions encountered in Africa. To be consistent with AIDSfreeAFRICA's focus on HIV and AIDS it is fitting that the drug is effective against oral thrush, a painful opportunistic infection that affects 20% of AIDS patients. Though the drug is produced in Ireland, is patented by the Belgium-based pharmaceutical company Tibotec, which makes the drug available to AIDSfreeAFRICA on a non-profit base. They use a non-profit distribution model specific for developing countries AIDSfreeAFRICA conducted a two-year pilot program to train doctors in the use of the drug. During that time the drug was made available free of charge. This program concluded successfully by creating the framework for AIDSfreeAFRICA, now authorized and licensed to sell the drug. Interestingly, as in most developing countries, Cameroon's government does not have the policies and regulatory agencies in place to handle an emergent pharmaceutical infrastructure. It took us 18 months to get the license and another 6 months to receive a printed copy of it! Imagine any pharmaceutical company starting production and then having to wait two years to get the authority to sell. Impossible, but true!

Cameroon needs more than pharmaceuticals and systems to distribute them. Hospitals, HIV/AIDS treatment centers, and diagnostic centers are all understaffed. Cameroon, like other countries in Africa, needs doctors, nurses and laboratory technicians. All of these experts are in short supply. Creating infrastructure to educate young people in these professions is also paramount. While the main focus of AIDSfreeAFRICA is the establishment of an infrastructure for pharmaceutical production, I will discuss some ways we are working to enroll others to help solve some of the related problems as well.

Cameroon's Education System

In order for AIDSfreeAFRICA to fulfill its mission of establishing pharmaceutical production in Sub-Saharan Africa, it has been necessary to assess the state of education. The factories need highly trained and specialized workers. Globalized chemical education will ultimately require that developing countries consider the critical nature of providing adequate scientific education–and that more developed countries educate scientists and science students about the needs of developing countries. AIDSfreeAFRICA aims to provide insights and connections to help increase awareness of the needs of developing countries, which will hopefully help alleviate the situation in less developed countries.

Let's begin where we all start schooling, in kindergarten. The Cameroonian schools can be split into three types; government-run, faith-based, and private. Children start school as young as three years of age, and attend three years of nursery school. Most children move on to primary school. Cameroon charges school fees which differ significantly according to the type of school. In general the school fees per year per student are US$8 for government-run public schools, attended by 57% of students in urban areas and 86% in rural areas. In government schools, teachers are underpaid and often not paid on time. The education is widely substandard. Faith-based schools charge US$30 and teach 12% of students (*18*) Private schools charge US$40 and operate almost exclusively in urban areas where they take 29% of the share of students (*18*). Faith-based schools are subsidized by the government, and are not necessarily better than private schools, which receive no government subsidies at all, but are overseen by the government.

Only 27% of students attend secondary school (32% of boys and 22% of girls). Secondary school in Cameroon is comparable to middle school in the US. Of these, 22% fail to pass the final exam and another 24% drop out. Those passing "O" level ("O" for ordinary) or "A" level ("A" for advanced) secondary (middle school) and upper secondary (high school), respectively, can go for training and be awarded diplomas in teaching, agriculture, nursing and several technical subjects. The trade schools prepare students for the job market rather than the University. After a good "A" level, a student applies to a University for undergraduate studies. The system is not consistent throughout the country and varies particularly between the Anglophone and Francophone provinces (*19*).

Students struggle to get a place in one of the six public universities: University of Buéa; University of Douala; University of Dschang; University of Ngaoundéré; University of Yaoundé I; University of Yaoundé II at Soa, or a place in one of five private Universities. A bachelor's degree is awarded after three to four years' of study at the university.

The minister of higher education makes final policy decisions regarding universities, although each university has a governing council that is responsible for personnel recruitment. The creation of new departments, degrees, courses and changes in regulations must receive ministerial consent. Each university receives a budget from the state.

The University of Buéa is headed by a vice-chancellor, nominated by the government, who in turn chairs the administrative council. Other public universities are headed by a rector. A Catholic University Institute was established in 1990. Several higher education institutions do not fall directly under the Ministry of Higher Education, but the Minister must ascertain that they meet academic standards. Some are run by other ministries and offer specialized training in agriculture, health, post and telecommunications, forestry and public works. Schools and Institutes in Administration, Technology, Social Work, and Public Works award diplomas in economics, management or law after two years. These are generally recognized as equivalent to an associate's degree in the US. A license can be obtained after three years in the Humanities, in the Sciences, or in Engineering. At Buéa, a bachelor's degree is awarded after three to four years' study. Students pursuing graduate programs must to go abroad. Many go to Nigeria to obtain higher education.

I had several occasions to visit the University in Buéa, which achieved its status in 1992, and features a well developed website. I would caution the reader when reading the website to keep in mind that in reality Cameroon is a resource-poor country, and perhaps this may need to be taken with a grain of salt (*20*). For my work in Cameroon I adopted former president Ronald Reagan quoting Vladimir Lenin: "trust, but verify." (*21*) Home to 12,000 students, until now the university offers dormitories only for first-year female students. A highly qualified and diversified staff of lecturers (300 permanent and 200 part-time) teach and undertake research. The University also employs about 473 support staff. The Chemistry department kindly showed me the classrooms and laboratories. Laboratory equipment seemed to be limited to analytical chemistry. Instrumentation, such as electronic pH meter, ion-selective electrodes, titration burettes, and some glassware were visible. Collaborations with European and US researchers are desired, pursued, and taking place.

A two-week long seminar called "Hands-on Research on Complex Systems," for example, is planned for this summer (August 2010). Among the half-dozen U.S universities involved is my alma mater, The City University of New York, and also New York University. According to the organizers, the two weeks will provide an interactive hands-on research experience involving tabletop experiments with real-time computer data acquisition and associated computational modeling. The research promises to be interdisciplinary and can be conducted by individuals or small groups using rather modest instrumentation. International faculty will lecture and lead small groups through the experiments and computer modeling. The Hands-on Research School has already conducted these seminars in Brazil in 2008, and in India 2009 (*22*).

The University has two libraries. The books are still catalogued with hand written index cards, but the librarian promised me that they are working on digitizing the system. There were a few computers. As I was to find out later, the university library was equipped with six computers and all with internet capability. The gap among universities in Cameroon became apparent to me when I visited the University of Yaoundé, built in 1962 in the capital of Cameroon (*23*). The university has two campuses, Yaoundé I hosts the sciences and Yaoundé II hosts the humanities. Being located in the capital and a much larger than the University in Buéa I may have expected too much, but nothing could have prepared me for the shock I had when I entered the universities main library. It was dusty, dingy, and in disarray. The handwritten card catalog showed that the books listed were from the 1970s and older. Eventually we were sent to a smaller library attached to the Department of Science and Medicine. This highly specialized library had a copy of the Merck Index from 1999 in French. The head librarian was overwhelmed by the promise of two copies of a current Merck Index and went immediately to inform the Chancellor of the University. Still in the library, I discovered six computers, all covered in plastic. I was told they were specifically hooked up to access the internet but since the internet was not working they were all covered up. I got a similar answer when visiting two chemistry professors who had just finished teaching. The professors gave me their private e-mails telling me that the school does not provide internet access.

The chemistry laboratory functioned as both lab and classroom. The lab benches all featured Buchi rotary evaporator equipment, and some glassware necessary to run an organic reaction. It was confirmed that the students had performed an esterification. As far as I understood from our conversation which was conducted in French and English at the same time, analysis is generally limited to thin layer chromatography. As far as I can tell, there is not a single nuclear magnetic resonance (NMR) machine in all of Cameroon. NMR machines are routinely used even by undergraduate students in the US for organic molecular structure elucidation and are invaluable for myriad research purposes.

I am very pleased that as a member of the International Activities Committee of the American Chemical Society, I have been asked to initiate collaborations between the US and African scientists. I am pleased to make my recommendations to the committee and facilitate to bring what is needed most, namely, equipment, chemicals and hands-on training courses on any donated equipment. A working session at the University of Yaoundé with the department chairs of chemistry, medicine and library, revealed what I had suspected, that sending US undergraduate students there in the hope that they could perform research in a laboratory and present their work in the form of a poster was not practical and not an appropriate first project, due to the lack of essential equipment, reagents and chemicals. A more suitable starting project might be for a professor to come with a few students and some basic laboratory equipment, such as an infrared and ultraviolet spectrophotometer, and spending time with students and professors alike, analyzing some of their natural extracts and other samples. Professors in Cameroon want support in educating the next generation of scientists in Africa. This new generation will be needed to fill the jobs opening up in the nascent pharmaceutical industry. This is indeed an exciting moment.

Overcoming Adversity - Training in Resource-Poor Areas

Despite these limited resources, it is it possible to train chemists well. The story will give the reader a glimpse into what it took to train a technician to make a diagnostic solution. A diagnostic solution is a reagent, in this case a water-based solution, containing salts, buffers and stabilizers. The reagent I was making functions to transport blood to be analyzed through a fine glass capillary where a laser beam detects the number and size of the different particles in the blood sample. A diagnostic reagent is not a drug, but it is used in a medical laboratory to analyze samples of body fluids, in this case blood. The final solution that AIDSfreeAFRICA wanted to produce is comprised of two inorganic salts and three organic additives, with a total concentration of dissolved solids of less than 1.5%.

Since AIDSfreeAFRICA focuses on drug production, the question arises, why it was important to train chemists to produce this diagnostic reagent. These diagnostic reagents play a crucial role in patient care. I learned this when I visited Kenya and Cameroon for the first time in 2005. At that time, Triomune was the drug of choice for first line treatment for HIV/AIDS. It is a three-drugs-in-one treatment, called the cocktail. The triple combination is designed to prevent drug resistance as much as possible. When resistance occurs or when a patient shows symptoms of side-effects, such as hepatic failure, the drugs have to be switched to second line treatment. In Cameroon, Triomune is provided and paid for by the Global Fund, and was readily available and dispensed for free. However, children's AIDS drugs and second line treatments are expensive and difficult to get. With first line treatment available and free to the patient, I did not understand why so many HIV positive patients were not placed on the drugs, although they met WHO guidelines for antiretroviral treatment. It turns out that patients are often simultaneously sick with either malaria, tuberculosis, fungal infections, and/or other infections for which drugs are often out of stock. I heard the term "out of stock" over and over in other hospitals. Even if the patient had the money to buy the prescribed medicine, it was not available because it was out of stock. AIDS patients could not take the available AIDS drugs because they were too sick with other illnesses for which the medicine was not available. Usually, the arrival of new supplies is unpredictable. "Just wait," the patients are told. Considering the distance and expense of travel for patients, if they go, they often fail to return. If they could afford the transport they would come back hoping the drugs had arrived.

Another reason why AIDS drugs sit on the shelves is that laboratory diagnostics are usually too expensive for the patients to afford. AIDS activists successfully fought and succeeded in securing availability of AIDS drugs and lowering drug prices, but not much was done to make diagnostics affordable. While a one month supply of Triomune cost US$6, full blood analysis costs US$32. The diagnostics were so expensive because each model of blood analyzer requires specific reagents and a considerable amount of liquid for the analysis. That reagent is typically composed of 98% water, two salts, buffer, stabilizer, and an antifungal agent. This simple solution is sold at full competitive market prices. In addition, the blood analyzer model used in Cameroon is out of date in the US; thus, while the reagent (specific for each model) was at that time produced for consumption only in developing countries, it was still fully priced. This situation can be remedied by making these reagents locally.

What if we could make the diagnostic reagent ourselves? Thus the solution was analyzed to find out the exact quantity and nature of the chemicals it was made off. Standard laboratory quantitative and qualitative analysis plus nuclear magnetic resonance spectroscopy was sufficient to find out the reagents composition. The next step was to find out where to buy the chemicals. I was able to find three of them in Cameroon, imported from Nigeria. I bought the remaining three in the US and took them with me on the airplane (*24*). I had only six days in Cameroon to train the technician, produce a test batch, and hand over production to the Africans. On the first day, I asked to see the lab. I wanted to make sure that all the chemicals had arrived. I had brought an analytical balance and a pH meter with a sufficiently sensitive pH electrode with me from the US. Indeed, everything had arrived in good condition. We were ready to start, but at that very moment the electricity went off. I waited for the noise that tells you a back-up generator kicks in. I am used to spending time in hospitals where these generators are a standard feature. However, because the production area was not part of a hospital unit, there was no backup generator!

The next day, my Cameroonian technician studied the procedure written by Dr. Elliott Bay, process chemist and member on the board of Directors of AIDSfreeAFRICA. This was no simple affair, even with my help. It became apparent that he did not know how to calculate molar concentrations, something I had hoped he would have learned in a first-year chemistry course. I showed him how to do it, and we began to make 25 liters of solution. There was no graduated cylinder, no balance with that capacity, and no container large enough to hold this amount of a liquid. We went to the market to buy a large container made of semi-translucent plastic. With a 1-liter measuring cup, we added five liters, one at a time, and marked the new container with masking tape. We did that five times for a total of 25 liters. I looked at the markings and found them somewhat uneven, although guessing was difficult since the container had a larger circumference at the top than at the bottom. I knew of a scale with 32-kg capacity, and I insisted we get special permission to carry the container into the sterile unit where it was located. We succeeded to put our solution on the balance to double check it, and thankfully realized that one of the five-liter units missed exactly one liter.

The plan was to produce the 20 liters of our diagnostic reagent by the third day. The technician analyzed the distilled water to make sure it was of the right quality; which meant he was titrating to check the hardness of the water. He then proceeded to prepare a stock solution of one of the two salts to make a series of solutions to establish a concentration curve. Sitting close to the tap water, I reminded him that he had to use distilled water to avoid contamination with other ions of unknown concentration. He looked up and innocently declared: "but Doctor, the distillation machine is broken and we have not ordered the new burn elements from the United States." That brought everything to a grinding halt. The supervisor of the water production unit had not been informed of the breakdown either. The technicians said that distilled water production was not scheduled for another month, so they wanted to wait. Upon hearing that, I asked if there was distilled water stored somewhere. Yes, there was, but analysis showed it to be contaminated. Most likely, the distillation was still in progress when the burners died, contaminating the only storage container.

When I am in Cameroon, I have three contigency plans for every task I am planning. In this case, I had to go to backup plan number three. I called a student from my collaborator's laboratory in Bamenda and asked him to go to the provincial hospital and buy 50 liters, or two containers, of distilled water. I knew the student was well trained and did not worry about his using a contaminated container or making a thoughtless mistake. Luckily he had the spare money to buy the water and the hospital distillation unit was working. I took an eight-hour night bus from Mutengene to Bamenda and got the water. Luckily, the next day I ran into another collaborator, Dr. Charles Boyo, a pharmacist and caught a ride back in a more comfortable vehicle. This left us two days for production and analysis. We were able to complete both without further incident.

In conclusion I must say that, however difficult it was, or however long it took to get the first 20 liters produced, in my experience, Africans work with precision, and are incredibly accurate. Batch after batch, the analysis came out with almost absolute precision. Producing this solution is one thing, however, but actually getting it into the hospital laboratories requires another long road filled with obstacles. Although we are producing a reagent twenty times cheaper than the import version, full-scale production has not been established. We are dealing with issues that range from finding a suitable factory to finding people who are willing to do something new to them. Furthermore, a diagnostic reagent must be registered in Cameroon as if it were a new drug application, requiring attention comparable to that required for FDA drug approval. In addition, the Minister for Public Health and the commissions dealing with these drug approvals are slow and not very responsive to inquiries concerning the approval process. Finally, because Cameroonians are so used to importing everything, they mistrust products made by their own citizens. To add to the complexity of the matter, there are taxation issues. It seems every time we climb one mountain, there is another in sight to overcome.

Concluding Remarks and Future Outlook

AIDSfreeAFRICA has big goals and these goals must evolve as we proceed. Take a for-profit pharmaceutical drug production and reframe it within the context and limitations of non-profit rules. Take an industrial chemist and drop him or her into a university laboratory in Africa. Take a program born in the air-conditioned offices of a pharmaceutical company whose employees have never been in a developing country and introduce that program to African pharmacists. Take an industrial chemist and turn him or her into a public relations person or a spokesperson for industrial development in underdeveloped countries. Take a volunteer brimming with "white people's" ideas and drop him or her into a village hospital without a centrifuge, microscope, diagnostic equipment, reagents, tests or blood storage facility. Take an African student who has never heard of chemical synthesis, and who thinks the only way to make drugs is by extracting plants and tree leaves. Then, somehow make these drugs available in a poor country where maybe half of the population can afford them.

From what I have seen in Cameroon over the last five years, it looks like it is time for a new paradigm of assistance. In general, most non-profit organizations are primarily concerned with education, which is an absolutely critical factor for success. But, too often, they focus on shipping donated goods that they think are needed, and hand them over with few or no requirements or feedback from the recipients. In fact, this type of charity, for all of its fine intentions, is only a momentary fix at best, but more damaging when planning for developing sustainable projects. The following story will hopefully demonstrate how the hope of free charitable money is foiling serious attempts to build long term sustainable development. When I first met Dr. Achu, I proposed finance micro-loan programs in the villages his organization was already working with. A possible requirement was that some the income generated be used to pay back the loan and then to pay for AIDS treatment related expenses of people who could not afford it. Dr. Achu turned to me and said, "If I have to pay the money back I don't want it. I will wait for the next white guy who gives it to me for free." It is regrettable, but this ended our beginning interest in working together.

Traditionally, philanthropic organizations have not established programs that are self-sustainable. Few such donations and efforts generated sustainable income. This is changing slowly however, as demonstrated by the popularity of the micro loan programs and organizations such as Kiva (*24*). According to Zambian native Dambisa Moyo, international economist and author of the *New York Times* best-selling book *Dead Aid: Why Aid is Not Working and How There is a Better Way for Africa*, we have a long way to go (*25*). Kofi Annan commented on the cover of the book that Dr. Moyo is too harsh in her critique on the role of aid, but agrees that "the determination of Africans, and genuine partnership between Africa and the rest of the world, is the basis for growth and development." We ought to consider moving on from simply donating goods and services to promoting full economic business development in partnership with the people we claim we wish to help. We ought to move more from giving fish to teaching how to fish, creating the infrastructure that allows the creation of jobs and the establishment of more than just a pharmaceutical company. This is why it is so important to connect to the universities in African countries and support efforts for student exchange and partnerships with universities across the Atlantic to provide books, laboratory equipment, teachers and professors. Educational opportunities supported by the basic resources needed to succeed can provide the basis for a continuum of job opportunities at home and a whole new prospect for sustainability.

AIDSfreeAFRICA wants to teach people how to fish, not just give them fish. Self-sustainability and sound business decisions must guide our work. When I visited Kenya I was introduced to a project that sends and replaces six medical doctors from Germany to the slums in Nairobi every eight weeks. The doctors come, deliver much needed medical care treating patients, and are saving countless lives. But they go back home with negligible training and development of the Kenyan medical staff. Sending doctors to the US for training is not practical either since the US, being short on doctors, is inviting foreign-born doctors to stay in the US. Much has been published concerning the "brain-drain" developing countries are suffering with their brightest people immigrating to developed countries (*26*). In fact, I have been approached by hospitals asking me to facilitate recruiting nurses from Cameroon, a country well known in the health industry for excellent nursing programs. Salaries, access to advanced technology, and comfortable lifestyles entice people to take up residence in the US, thus further depleting Cameroon of urgently needed medical experts. AIDSfreeAFRICA seeks to create sustainable programs that can flourish in Cameroon.

There are plenty of donated generators, motor bikes, and even autoclaves (sterilization equipment), but no financial source to keep them going, and no technical expertise to keep them maintained and fixed when broken. Chemical and technical education and an income-generating activity that serves as a dedicated funding source to keep these projects going is a must if we want to avoid littering the African landscape with broken down and abandoned equipment, some of them oozing fuel, lubricants and other chemical hazards.

The Prime Minister of Cameroon, Philemon Yang, received AIDSfreeAFRICA enthusiastically. I hope the government will support our efforts by providing licenses and inspections in a timely fashion. We also hope to be able to negotiate generous tax agreements on imports of raw materials and production equipment. The design of AIDSfreeAFRICA, and the plan describing how the organization aims to solve these problems was recognized in the Buckminster Fuller Architectural Design Contest in 2009, where it advanced to the semifinal stage. As founder of AIDSfreeAFRICA, I was awarded my first humanitarian award, a US$30,000 prize from the US Astellas Foundation, and awarded in an American Chemical Society Presidential Event hosted in conjunction with the American Chemical Society meeting in Washington, D.C. in August, 2009.

In Africa, there is no size that fits all. Solutions will vary significantly from one African region to another, more so than is currently practiced in pharmaceutical production in the US and other developed countries. Production in Africa will have to look and feel "African" if it is to succeed in a sustainable way.

As I do my last editing, I am back again in Cameroon. I welcome and appreciate the very latest development. To understand the significance, one must appreciate that in 2005 India joined the World Trade Organization. To be admitted to this distinguished club of countries, India had to promise to recognize international patent laws. As soon as that was announced, prices for drugs and drug production in India went up. Indian manufacturer CIPLA has been looking for less expensive places to move to and expand their drug production. It found one partner in Uganda, which, to the best of our understanding, and after three years, has not yet succeeded in starting production. CIPLA, however, backed up by the German Bank, made a similar arrangement to partner with Cinpharm, which is owned and operated in Cameroon. Yesterday, I had the privilege of inspecting the factory and meeting the people in charge.

I am pleased. The facility is applying for WHO standard approval and may only be a few months away from production. It will now be the task of AIDSfreeAFRICA to monitor, troubleshoot, expand and support this effort with expertise and encouragement. Africa's emerging pharmaceutical industry is seeing the light at the end of the tunnel. Congratulations to everyone who has persevered.

References

1. *The World Medicines Situation*; World Health Organization: Geneva, Switezerland, 2004. http://www.globalissues.org/issue/2/causes-of-poverty (accessed June 28, 2010).
2. Medicines supply in Africa. *BMJ* **2005**, *331*, 709. http://www.bmj.com/cgi/reprint/331/7519/709.pdf (accessed June 28, 2010).
3. Grand Challenges in Global Health. http://www.grandchallenges.org/ (accessed May 22, 2010).
4. Dagani, R. Gates Foundation awards 43 grants for global health research. *Chem. Eng. News.* **2005**, *83,* 9. http://pubs.acs.org/cen/news/83/i27/8327notw9.html (accessed June 1, 2010).
5. Teixeira, P. R.; Vitória, M. A. A.; Barcarolo, J. The Brazilian Experience in Providing Universal Access to Antiretroviral Therapy. In *Economics of AIDS and Access to HIV/AIDS Care in Developing Countries: Issues and Challenges*; Moatti, J.-P., Coriat, B., Souteyrand, Y., Barnett, T., Dumoulin, J., Flori, Y.-A., Eds.; Agence Nationale de Recherches sur le Sida: Paris, 2003; pp 69–88. http://www.lepublieur.com/.
6. Brazilian STD/AIDS Programme, Brasilia. *Epidemiol. Bull.* **2006**, III, Epidemiological weeks 1–26. www.aids.gov.br.

7. *Towards Universal Access: Scaling Up Priority HIV/AIDS Interventions in the Health Sector*, 2008. World Health Organization. http://www.who.int/hiv/pub/2008progressreport/en/ (accessed on June 1, 2010).
8. *Resposta Positiva: A Experiencia do Programa Brasileiro de AIDS*; Ministry of Health of Brazil: Brazil, 2002.
9. Wise, J. Access to AIDS medicines stumbles on trade rules. *Bull. W. H. O.* **2006**, *84*(5), 337–424. http://www.who.int/bulletin/volumes/84/5/news10506/en/index.html.
10. HIV and AIDS in Brazil. AVERT. http://www.avert.org/aids-brazil.htm (accessed June 1, 2010).
11. CIPLA–Caring for Life. www.cipla.com (accessed June 1, 2010).
12. Increasing Access to Essential Drugs. Health Action Internation. http://www.haiweb.org/pdf/FocusAreas_Access_E1.pdf (accessed June 1, 2010).
13. Cameroon Statistics. UNICEF. http://www.unicef.org/infobycountry/cameroon_statistics.html#67 (accessed June 1, 2010).
14. Schaefer, P. The Seven Pitfalls of Business Failure. Attard Communications. http://www.businessknowhow.com/startup/business-failure.htm (accessed June 1, 2010).
15. Institutions. Cameroon Baptist Convention Health Board. http://www.cbchealthservices.org/html/our%20institutions.html (accessed June 1, 2010).
16. ISO9001 is a national standards measurement commonly used by industry and other organizations. For more information visit http://www.iso.org/.
17. Miconazole. Wikipedia. http://en.wikipedia.org/wiki/Miconazole (accessed June 28, 2010).
18. Backiny-Yetna, P.; Wodon, Q. *Comparing the Private Cost of Education at Public, Private, and Faith-Based Schools in Cameroon*; Munich Personal RePEc Archive (MPAR) Paper 16464, 2007 data; Munich University Library, Germany, 2009.
19. Cameroon: Secondary Education. StateUniversity.com. http://education.stateuniversity.com/pages/230/Cameroon-SECONDARY-EDUCATION.html (accessed June 1, 2010).
20. University of Buéa. http://www.ubuea.net// (accessed June 1, 2010).
21. Trust, but Verify. Wikipedia. http://en.wikipedia.org/wiki/Trust,_but_verify (accessed June 28, 2010).
22. Foba, J.; Ngwa, G.; Sreenivasan, R.; Roy, R.; Showalter, K.; Swinney, H. Hands-On Research in Complex Systems School. The Abdus Salam International Centre for Theoretical Physics. http://www.handsonresearch.org/ (accessed June 1, 2010).
23. A nice gesture from the manufacturer was slapping on a big label reading, “Safe for transport in an airplane.”
24. KIVA is a microloan program meant to provide small loans to people in countries with an undeveloped banking structure. More information can be found at www.kiva.org.
25. Dambisa, M. *Dead AID: Why Aid is Not Working and How There Is a Better Way for Africa*; Farrar, Straus, and Giroux: New York, 2009.

26. Mutume, G. Reversing Africa's "brain drain": New initiatives tap skills of African expatriates. *Africa Recovery*. July **2003**, *17*(2), 1.

Indexes

Author Index

Subject Index

Printed and bound by CPI Group (UK) Ltd, Croydon, CR0 4YY